# Management of
# Construction Projects

# Management of Construction Projects

**Brian Cooke, MSc**

WILEY Blackwell

This edition first published 2015
© 2015 by John Wiley & Sons, Ltd

*Registered Office*
John Wiley & Sons, Ltd, The Atrium, Southern Gate, Chichester, West Sussex, PO19 8SQ, United Kingdom.

*Editorial Offices*
9600 Garsington Road, Oxford, OX4 2DQ, United Kingdom.
The Atrium, Southern Gate, Chichester, West Sussex, PO19 8SQ, United Kingdom.

For details of our global editorial offices, for customer services and for information about how to apply
for permission to reuse the copyright material in this book please see our website at www.wiley.com/
wiley-blackwell.

*Library of Congress Cataloging-in-Publication Data*

Cooke, B. (Brian)

Management of construction projects / Brian Cooke.
     pages   cm
    Includes index.
    ISBN 978-1-118-55516-3 (pbk.)
1.  Building–Superintendence.   2.  Building–Case studies.   I. Title.
   TH438.C6456 2013
   690.068–dc23
                                      2013024689

A catalogue record for this book is available from the British Library.

ISBN: 9781118555163

Wiley also publishes its books in a variety of electronic formats. Some content that appears in print may not
be available in electronic books.

Cover image: © iStockphoto / Marcusphoto1 (front cover, leftmost image in the top row)
Cover design by Steve Thompson

Set in 9/12.5pt Myriad by SPi Publisher Services, Pondicherry, India

1   2015

# Contents

Preface    vi

Acknowledgements    viii

1   **Organisation of the Construction Process**    1

2   **Developing Construction Teams**    24

3   **Project Planning – Linked Bar Charts and Procurement Programmes**    34

4   **Contracts**    44

5   **Materials Management**    55

6   **Mechanical Handling and Risk Assessment**    73

7   **Managing Construction Defects**    95

8   **Hotel and Office Project Development**    105

9   **The Merlin Project**    143

10   **The Co-operative Head Office Building**    165

11   **Chinley School Project**    195

12   **Retail Unit and Car Park**    222

13   **University Refurbishment Project**    251

14   **Managing a Small Business**    272

Index    293

# Preface

Managing Construction Projects is based on personal observations of six construction projects, with the text following the work stages from commencement to completion of the projects. The content is based on the author's observations of various aspects of each project. The information collected was freely given by the project managers, site managers and surveying staff based on the projects.

The head office and site organisation structure are outlined on each of the projects in relation to the main contractor. Relationships between the client's and the contractor's team were also considered.

Site planning and programming routines on the various contracts are outlined. Extensive use was evident on all the projects that Asta Developments – Power Project planning software was extensively used. All projects relied on laptop computers to provide access to company documentation and procedures.

Observations of the methods of handling a wide range of components – including large storey height panels, precast coffered floor units and curtain wall panels are illustrated. Site layout plans are also illustrated, together with examples of material storage proposals. Good and bad materials management practices on both the large and smaller projects are also discussed.

A number of construction sequences are shown for the erection of a 15 storey steel frame, 10 storey pre-cast cross wall frame and related buildings. The importance of construction method statements is dealt with in relation to the concreting of a large in-situ concrete floor.

Separate chapters are included to supplement the case studies with additional notes and examples. These include:

- Project organisational responsibilities based on an understanding of the knowledge and management skills required to undertake the various site management roles on a project
- The principle of developing a team approach for managing future projects is indicated. This relates specifically to developing site management, quantity surveying and site engineering teams which can be transferred from project to project.
- Programming and planning projects, based on the Power Project linked bar chart software is illustrated for setting up a bar chart display. The relationships between the operations are shown together with the principles of establishing procurement displays.
- Risk assessment displays are shown for a range of site based tasks involving major plant items.
- The importance of understanding the range of JCT contracts available in practice is an essential requirement for both the project manager and contract surveyor – these are summarised from information available from the JCT.

It is hoped that students and site management personnel obtain as much benefit from the material as I have enjoyed in preparing it.

Managing Construction Projects will be of interest to both lecturers and students on the following College and University courses:

BTEC UNITS – Taught subject areas included in the programmes of study include: Law and Contracts, Health Safety and Welfare, Project Management (Planning and organising), Management Principles (The role and responsibilities of site management personnel)

BTEC LEVEL 5 HND in Construction and the Built Environment – Taught subjects include: Site Surveying Procedures and Practice, Technology of Complex Buildings, Project Management (Roles and Responsibilities, Organisation Structures and Team Building)

BSc PROGRAMMES – Degree Courses in Construction Management / Quantity Surveying and related subject areas at level 4, 5 and 6. These include related subjects of Procurement and Project Information, Construction and Site practice, Commercial Management, Contract Practice and Site Production Studies

Many degree programmes incorporate group project work which involves team building and analysing construction sequences. Tasks also include presenting construction programmes, the selection and use of plant, safety practices and materials management and site layout planning. This text book would assist students in improving their presentation skills when preparing joint or individual coursework or projects at the final year of a degree programme.

# Acknowledgements

This book could not have been written without the dedication and assistance from a team of colleagues I wish to thank. These include:

- Paul Hodgkinson – We have worked together as a team on my last three books and once again Paul has given one hundred percent commitment – even during illness in latter stages of preparing the book.
- Sarah Peace – a delight to work with.
- The team at John Wiley & Sons.

One cannot thank enough the company personnel who freely contributed to the case study material. This includes site and head office personnel at BAM, Galliford-Try, Goyt Construction, Mansell, Morgan-Sindall, Pochin Construction and Wates Construction.

Brian Cooke (July 2014)

# Chapter 1

# Organisation of the Construction Process

## Contents

| | | |
|---|---|---|
| 1.1 | Overview of the size of the companies included in the case studies | 2 |
| 1.2 | Approach to the management of projects included in the case studies | 3 |
| 1.3 | Organisation principles applied to construction firms | 4 |
| 1.4 | Functional relationships and line management | 5 |
| 1.5 | Roles and responsibilities of site management personnel | 6 |
| 1.6 | Background experience and qualifications for construction personnel | 8 |
| 1.7 | The project manager | 9 |
| 1.8 | The site manager | 14 |
| 1.9 | The planning engineer | 15 |
| 1.10 | The project surveyor | 16 |
| 1.11 | The procurement manager | 16 |
| 1.12 | The site engineer | 18 |
| 1.13 | The clerk of works | 21 |

*Management of Construction Projects*, First Edition. Brian Cooke.
© 2015 John Wiley & Sons, Ltd. Published 2015 by John Wiley & Sons, Ltd.

## 1.1 Overview of the size of the companies included in the case studies

The range of construction firms related to the case studies have been categorised as follows, with respect to organisation size.

- Small firm: 1–49 direct employed staff and operatives
- Medium firm: 50–299
- Large firm: 300–1199
- "Big" firm: over 1200

Project 1: Hotel and office – Galliford Try
Project 2: Industrial factory – Pochin Construction
Project 3: Co-operative office – BAM Projects
Project 4: School project – Mansell (Balfour Group)
Project 5: Retail unit / Car Park – Morgan Sindall
Project 6: University refurbishment – Wates Construction
Project 7: Housing project – G. Construction

| Company | Direct employees | Project value £M | Size/category |
|---|---|---|---|
| Galliford Try | 2500 | 12.0 | Big |
| Pochin | 250 | 14.0 | Medium |
| BAM | 2500 | 117.0 | Big |
| Mansell | 1000 | 4.0 | Large |
| Morgan Sindall | 4000 | 11.0 | Big |
| Wates Construction | 3500 | 12.0 | Big |
| G. Construction | 10 | 1.0 | Small |

## 1.2 Approach to the management of projects included in the case studies

The majority of organisations in the project case studies undertook a functional approach to the management of their projects. BAM, however, indicated in their company data that they have adopted a matrix organisational structure for the management of projects (see later notes on matrix management).

On contracts up to £5 M in value the project manager / site manager was responsible for direct control of the project. They were supported by visiting personnel undertaking the functions of quantity surveying / planning / design team co-ordination and safety management.

On the larger projects, over £10 M in value, all these functions were site based.

A site organisational structure is indicated for each of the projects in the case studies.

The number of permanent site staff is shown, together with the number of visiting personnel.

It is common practice to place a planning engineer and design team co-ordinator on a major project and allow them to service additional smaller projects from the same contract base.

The site planning engineer had often been involved in the project from the tendering stage. Planning responsibilities often include preparation of:

- pre-tender programme
- contract programme
- procurement programme
- programme progress updates during construction
- programming to completion.

In all the programming stages of a contact Power Project or Team Plan was the commercial software package used. Separate notes are included on using linked bar chart software in practice.

In all the case studies, the site based quantity surveyors were under the direct control of a commercial manager based at head office.

On the larger projects, over £10M, a senior project surveyor and up to two assistant surveyors were engaged on site. The surveying functions undertaken by the team included:

- liaison with the design team co-ordinator
- preparing monthly payment applications
- dealing with variations to contract
- payment to work package contractors
- preparing cost/value reports for senior management (which took at least 10 days per month to report and finalise).

It was noted that project managers were directly involved in the cost/value reconciliation process at each month end. They considered that cost/value analysis was simply a paper exercise to warrant the surveyor's existence. The managers were fully aware that the surveying team would "produce the white rabbit out of the hat" at the appropriate time to save the contract situation. How true this is, from an observer's position!

Headings included on the organisation of the construction process include:

- organisation principles applied to construction firms
- functional relationships and line management
- roles and responsibilities of site management personnel including:
    project manager
    site manager
    planning engineer
    project surveyor
    design team co-ordinator
    site engineer
    clerk of works

In assessing their roles, consideration has been given to the knowledge requirements and management skills necessary to perform their job successfully.

## 1.3   Organisation principles applied to construction firms

Cole's *Management Theory and Practice* summarises common forms of organisation structure as being:

- functional organisations
- product based organisations
- geographical/or regional based
- divisional organisations – based on product or regional and having key functions reserved for head office
- matrix organisational structures – see separate example in Section 2.2.

Construction firms often fall into a combination of divisional/regional organisations (with one central head office co-ordinating the regional organisations).

Companies generally operate on a functional basis.

The head office undertakes the following functions, which give support to the various projects:

- estimating (estimating and tendering)
- surveying functions
- administration services
- health and safety function
- human resources services
- contracts (including the planning function)

Construction firms fall into four categories according to the number of direct employees. Government statistics indicate that ninety per cent of firms in the UK fall into the small category (1–49).

An interesting question to pose is to attempt to identify the number of "big construction firms" in your region of the country. Try to identify **ten** construction firms.

For example, large companies in the North West include:

Laing O'Rouke, Taylor Wimpey, Wates Construction, Bovis, Balfour Beattie, Carillion, Morgan Sindall, Robert MacAlpine.

## 1.4   Functional relationships and line management

The organisation of a major project is based on functional relationships. Line management allows direct authority over others, which is the essence of a "chain of command" during a construction project. Illustrations and information are passed down the chain and responses communicated back up the chain. Line management provides a two-way communication system.

Examples of line management are illustrated for the site management and surveying functions.

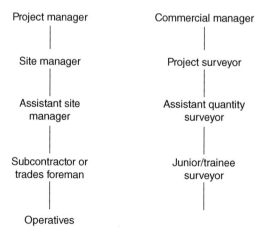

**Line management**

A construction project is based on controlling site functions such as:

● construction function
● surveying function
● design team function
● planning function.

It is the project manager's role to pull these functions together and develop a competent project team. The development of a united team spirit will often lead to a successful project. Team building within a construction company is essential for the continued success of the business.

Wates Construction aims to develop a team approach to serve specific types of projects and clients.

## 1.5 Roles and responsibilities of site management personnel

The organisation structure for a £12 M building refurbishment project is shown here. The roles and responsibilities of various site personnel are outlined separately in this section. This will also include the role of the clerk of works. The

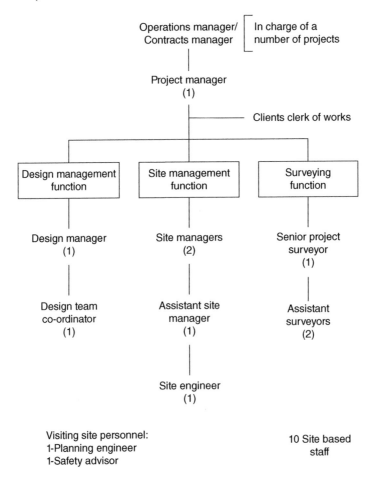

Operations manager/ Contracts manager — In charge of a number of projects

Project manager (1)

Clients clerk of works

Design management function

Site management function

Surveying function

Design manager (1)

Site managers (2)

Senior project surveyor (1)

Design team co-ordinator (1)

Assistant site manager (1)

Assistant surveyors (2)

Site engineer (1)

Visiting site personnel:
1-Planning engineer
1-Safety advisor

10 Site based staff

three main functions illustrated are design team, site management and surveying. Other functions, such as planning and safety, are provided by visiting site personnel.

The roles and responsibilities of a range of site management personnel are now outlined.

## Organisation structure of a regional contracting organisation

This large organisation is a family-owned business with direct involvement at senior management level. The group incorporates nine regional offices in the UK, including the Midlands, north-west England, Yorkshire and the north-east. The company head office is located in London and has up to 3500 directly employed personnel.

### Regional organisation – north-west England

The north-west region operates in three areas of construction activity: main contracting, housing and refurbishment and retailing and interior fit-outs.

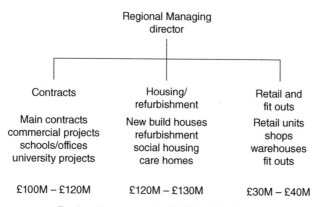

Regional Managing director

| Contracts | Housing/ refurbishment | Retail and fit outs |
|---|---|---|
| Main contracts commercial projects schools/offices university projects | New build houses refurbishment social housing care homes | Retail units shops warehouses fit outs |
| £100M – £120M | £120M – £130M | £30M – £40M |

Regional turnover in the £250M – £300M range

**North-west region – overall divisions**

### Technical services across the region

**Estimating and tendering** – This is provided within the region as a joint service to each of the three divisions. This is under the control of bid centre manager, and is aimed at tailoring the service to each of the market areas.

**Surveying functions** – This is under the control of the business commercial manager who allocate an experienced surveying team to each of the specialist areas (main contracts, refurbishment and retail projects).

Likewise the procurement function is managed by a regional procurement manager.

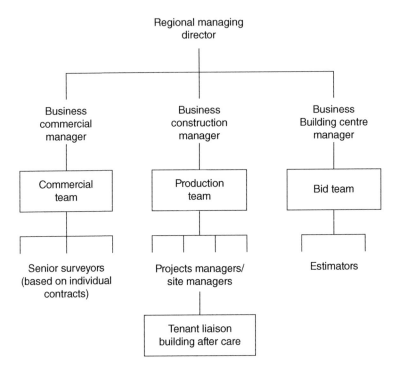

**North-west region management structure – refurbishment division**

The business construction manager mirrors the role of a contracts manager in a similar organisation (i.e. a manager in charge of a number of contracts or managed by project management personnel.)

The tenant liaison team is necessary on refurbishment projects to liaise with tenants or occupiers and deal with building aftercare, i.e. teething problems after tenants occupy the premises.

**Note**: This case study is an interesting approach to the management structure of a large contracting organisation operating in three different construction fields within a single region.

The overriding feeling is that it successfully works.

## 1.6 Background experience and qualifications for construction personnel

### Contracts/operations manager

The post of contracts manager is an esteemed position in a contractor's organisation. Dealing fairly with people is an essential feature of their character. Communication skills with clients, senior project managers and site management personnel is essential, but to be liked by **all** is not a trait to admire. Sometimes one has to be feared or disliked to hold the respect of the management.

## Project managers

In a construction environment, experience is often more highly regarded than paper qualifications. Routes are available for site and office personnel to join professional institutions such as the Chartered Institute of Building without having studied at a university; this is via the experienced practitioners' routes to membership (MCIOB). Graduate entry is also available after a minimum of three years' experience. Project managers often earn their positions by proof of efficient management on similar value projects. The development of team building is essential for their success in managing projects. Good project managers create construction teams that they carry from one contract to another.

## Site managers

Site managers and assistant site managers often have a trade background. "New" site managers with a degree background are usually placed on an in-company training programme. The training programme usually covers a three-year period, after which the manager can apply for MCIOB status (Chartered Building Status).

## Site engineers

Many site managers commence their site experience as a site engineer. Background study courses often include an HNC/HND qualification. On completion of a two- or three-year training period, promotion to assistant site manager would be considered the norm.

## Project planning engineer

In the role of assistant site manager or site manager, experience is gained in preparing programmes, monitoring progress and writing report. The introduction of computer software based on linked bar charts has revolutionised planning and programming at site level. A manager showing a flair for developing programming skills may ultimately result in the person becoming a planning engineer. An understanding of construction sequences is an essential requirement.

We will now look at the roles and responsibilities of the various levels of management.

## 1.7  The project manager

## Knowledge requirements

- To be familiar with company procedures
- To be familiar with all aspects of the construction process in respect to the management of a project
- To understand the key requirements of the project: planning, controlling and reporting to the contract operations/contracts manager

- To understand the responsibilities to the client or the client's design team and site management personnel
- To have an understanding of the form of contract with respect to the impact of variations, possession, extensions of time and dispute resolution
- To understand procedures when dealing with disputes, especially with regard to subcontractors and suppliers

## Management skills

- To establish leadership skills when dealing with site management and subcontractor representatives
- To develop a team approach among their site management personnel
- To delegate responsibility to the site management team
- To maintain good site records when reporting on progress and on the contract profitability situation to senior management
- To be proficient in report writing and communications with senior management, site staff, subcontractors and the client
- To implement company procedures and policies
- To act as mentor to immediately subordinate management personnel

When considering management skills one must consider the application of the seven principles of management: leadership, delegation, organising, communicating, planning, forecasting and control.

## Around the project manager's office

**Permit board where site personnel can collect daily permits. Display area showing site photographs file and visitors' information.**

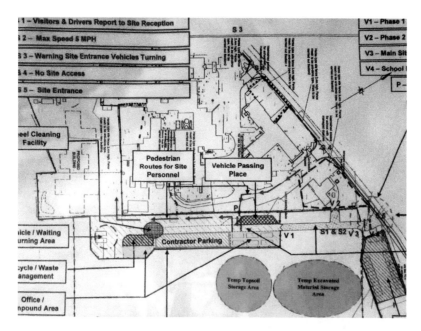

Site layout plan with key access areas and traffic movement areas. Material storage areas shown.

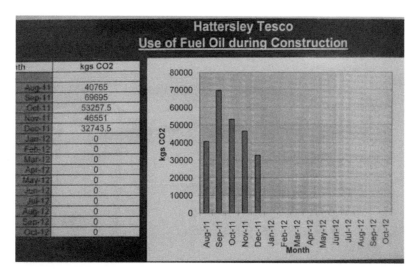

Environmental display board showing planned and actual water used, electricity and waste management records.

External site progress and photograph area. This is a common feature for a project. Good file management is a necessary requirement of a well-managed office. The use of a daily whiteboard to highlight key daily tasks.

Office filing system

Daily task whiteboard

## DAILY TASK BOARD

| MON | TUE | WED | THU | FRI | SAT |
|---|---|---|---|---|---|
| Services cut outs - B1.2 | Contact P/L S/C | Monthly site meeting | Start drain connection block C | Arrange Tr. lights for monday | Plasterers working |
| P.M. senior staff meeting | Architect V.Os schedule | | | Const. director's visit | |
| Brick delivery sch. required | | | | Building control visit/p.m. | |

## DAILY HAZARD BOARD

| | |
|---|---|
| MON | Ready mixed concrete wagons adjacent to lift shaft<br>New safety barriers erected<br>Dismantle south scaffold access tower block B |
| TUE | Mobile crane unloading steelwork - rear elevation<br>Pedestrian access route changed at S.W. corner of site |
| WED | Moving plant/lorry access to basement area |
| THU | Unloading table forms in basement area - moving plant |
| FRI | |

| NAME AND SHAME BOARD | | | |
|---|---|---|---|
| **DATE** | **LOCATION** | **OCCURANCE** | **SUBCONTRACTOR** |
| Tue 15th Sept | Drainage work MH 7–9 (rear block 3) | Inadequate support to drainage trenches<br><br>Barriers failed to be erected at end of day | Evans drainage company |
| Wed 16th Sept | First floor landing & corridor areas | Work area left in an uncleared state - timber off cuts/ loose timber | B & R joinery |
| Wed 16th Sept | Perimeter scaffold 2nd floor front elevation | Missing toe boards and handrails to staircase exit (area to be cordoned off) - untill dealt with | HR scaffolding |

## 1.8 The site manager

### Knowledge requirements

To be familiar with site procedures.

- To deal with site induction process for own labour and subcontractors and visitors
- To deal with requirements outlined on method statements and risk assessments prior to commencing operations on site
- To keep and maintain site records with respect to progress, contract delays and variations to contract
- To understand contact procedures regarding the issue of permits
- To be computer literate with respect to computerised processes to aid the construction manager – extensive company procedures are transferable from the companies website, e.g. health and safety policy, method statements and risk assessment formats
- To understand basic programme techniques including progress recording

### Management skills

- To develop communication skills with operatives and subcontractors when issuing instructions
- To show evidence of computer competency in order to maintain site records and access company procedures
- To show competency in the organisation of subcontractors

- To participate in the planning of work for future site operations, e.g. short-term planning procedures adopted by the company
- To develop leadership skills with reference to assistant site managers and management trainees
- To develop motivation skills by encouraging progression to higher management levels

Site managers are recruited from a wide range of backgrounds. Many will have developed organisational skills at general foreman or supervisor level. Management skills may have been developed by mentoring and by attending in-company training programmes.

## 1.9  The planning engineer

### Knowledge requirements

To develop an understanding of construction sequences and the order of work for various forms of construction, such as concrete frames of steel frames.

- To have an understanding of the various planning and programming techniques used by contractors
- To be familiar with the company's planning software
- To understand the various stages of planning during a project: pre-tender, pre-contract and contract planning stages
- To understand the link between programming and procurement programmes, with respect to work package subcontractors

**Note**: Planning Engineers normally attend internal company training courses or courses provided by the software specialists, such as Power Project or Asta Developments.

### Management skills

- To have good presentation skills
- To assist in the training of others in programming and presentation techniques, including mentoring assistant or trainee planners
- To prepare concise weekly and monthly reports on the contract progress situation
- To have good communication skills with all levels of the management team
- To liaise with the pre-contract team in the programming of work package subcontractors
- To contribute to the meeting when considering project acceleration

The planning engineer may be supervised by a senior planner, and may be responsible for planning and updating programmes on two or more projects . Many planning departments contain a number of pre-tender planners (dealing with programmes for tenders). Contract planners tend to be located on major large projects and have often been involved in the project since the tender stage.

## 1.10 The project surveyor

This person is one who has held a similar position on a previous contract. They must be familiar with all aspects of surveying from the tender stage through to the settlement of the final account. Their direct supervisor is the regional commercial manager to whom they are directly responsible, and they are also responsible for reporting directly to the project manager.

## Knowledge requirements

- To understand the company's tender, pre-contract and contractual process regarding the appointment of subcontractors and work packages
- To understand the valuation and payment process regarding package contractors and domestic subcontractors
- To be familiar with the company's cost/value reporting process to senior management
- To understand the monthly valuation process
- To be conversant with the form of contract in respect of payment and dispute resolution
- To be familiar with work package subcontractor procedures
- To be familiar with the preparation of dates to subcontractors for an extension of time

## Management skills

- To delegate to, and control, the site surveying team
- To co-ordinate with the procurement managers regarding work package contractors and domestic subcontractors
- To provide cost advice to estimators and managers at the tender preparation stage
- To manage the cost/value reconciliation (CVR) process on site
- To report to the project manager and commercial manager on the cost/value position at the end of each month
- To assist in the training programme of their assistant surveyor
- To prepare monthly valuations and manage payments to work package subcontractors

## 1.11 The procurement manager

The procurement manager may fall under the direction of the project surveyor due to the latter's direct link with the work package subcontractors. As an alternative they may report direct to the project manager.

## Knowledge requirements

A knowledge and understanding of

* materials and product availability in the construction market
* types and forms of subcontract
* the interface between work packages
* the tender/bid process
* financial terminology
* warranties, bonds and provisional suns
* the planning process and planning software used by the company
* contract programmes and the effect of delays on work packages

## Management skills

* To steer project mangers and commercial staff to correct solutions with regards to work package problems
* To visualise the bigger project picture without getting lost in the detail
* To understand how to analyse market conditions including the marketing business
* To understand business plans
* To negotiate with respect to management skills, tactics and price
* To report to senior management on marketing matters and market trends

The role of the procurement manger differs between the tender, pre-contract and contract stages.

At the tender stage:

* Selection of suppliers and subcontractors to be invited to tender
* Analysis of bids and recommendations to bid manager on suitable subcontractors
* Negotiation with subcontractors to obtain savings and reduce risk
* Evaluation of value engineering options – proposals to reduce price by joint discussion and agreement
* Risk evaluation on the selected subcontractors
* Manage the mid-tender interviews with selected subcontractors
* Check subcontractors' capacity to deliver the packages – consider areas such as current workload
* Consideration to widen the current subcontractors lists and introduce new subcontractors onto tender lists

At the pre-contact stage:

* Final negotiations on cost and value engineering with each subcontract package
* Final subcontract interviews – visit subcontractors' offices and other current projects
* Handover of bid information to operations team, and ensure that any savings and value engineering options are understood by the delivery team

At the contract stage:

- Ensure that all subcontractors are approved by the company
- Promote the company buying code to all subcontractors
- Monitor and track the spending on rebates
- Review and sign off the orders before issuing them to the subcontractor
- Undertake regional overview of subcontractor's performance, and measure and record the key performance index matrix
- Continual review of the robustness of the subcontractor, in terms of financial stability
- Develop the supply chain into new subcontract areas
- Promote the company at "meet the contractor" events throughout the region

## 1.12   The site engineer

### Knowledge requirements

- An understanding of the use and site applications of modern surveying equipment and methods, e.g. electronic distance measuring equipment (EDMs)
- Familiarity with the use of laser levelling equipment
- An ability to read and understand construction drawings
- An understanding of construction sequences: the order of work
- A basic knowledge of construction technology to at least B.Tec or HNC level

### Management skills

- To communicate with trades foremen and operatives alike, with respect to setting out procedures
- To maintain accurate site records of setting out procedures by recording data on site layout drawings and in dimension books
- To work as a team with the site manager or foreman and subcontractors on setting out procedures
- To develop organisational skills in the supervision of subcontractors and trade gangs
- To assist in the training of assistant or junior engineers

Many site management trainees commence their site experiences as an engineer or engineer's assistant. The theory and skills they have been taught at a university or technical institution now become a reality – and what a cultural shock it can be.

## The site engineer's role

Engineers should be competent at using a wide variety of surveying instruments including those illustrated.

They should understand co-ordinate surveying, which is now the main method of accurate setting out on site.

Familiarity with the engineer's level and laser levelling equipment is necessary. This is in order that he can instruct working foremen how to use the equipment.

They must be capable of keeping the level and dimension book up to date. Their responsibilities often include the training of young inexperienced engineers.

**DeWALT laser level**

**Foreman checking level of foundation concrete/ checking compact stone fill levels using the laser level**

**Engineer and foreman setting out pile position from string lines**

**Freelance engineer setting out on virgin site**

## Typical surveying equipment

A range of surveying equipment is illustrated. The site engineer must be fully conversant with using the surveyor's electronic level, laser level, theodolite and distance measuring equipment.

**A Pentax laser level in use for foundation work**

**An original three screw dumpy level**

**A Leica distance bearing theodolite**

**A ball and socket quickset level – a Swiss-made Kern instrument**

**A levelling metric staff**

## 1.13   The clerk of works

These notes have been abstracted from a publication by the Institute of Clerk of Works and Construction Inspectorate – version 2 March 2010.

### Main responsibility of the clerk of works

Their main responsibility is to make sure that work is carried out to the client's standards, specification and schedule. In most cases, the specifications are prepared by architects or engineers employed by the client. The clerk of works makes sure that the correct materials and workmanship are used and that the client is given quality work and value for money.

The clerk of works is either on site all the time or at least makes regular visits. They need to be vigilant in their inspections of a large range of technical aspects of the work. This involves:

* becoming familiar with all the relevant drawings and written instructions, checking them and using them as a reference when inspecting the work
* making visual inspections
* taking measurements and samples on site to make sure that the work and the materials meet the specifications and quality standards
* being familiar with legal requirements and checking that the work complies with them
* having a working knowledge of health and safety legislation and highlighting shortfalls observed to the person(s) concerned.

Clerks of works are not only inspectors, but also superintendents. This means that they can advise the contractor about certain aspects of the work, particularly if something has gone wrong. They can also agree to minor changes. However, they cannot give advice that could be interpreted as an instruction, particularly if this would lead to additional expense. Any verbal instructions must be confirmed by the architect.

They keep detailed records of various aspects of the work, which they put together in regular reports to the architect or planner and to the client. Records include details of:

* progress and any delays
* the number and type of workers employed
* weather conditions.

## Around the clerk of works office

The clerk of works office is combined with the materials sample display area

Full-scale displays are often provided to indicate the cladding finishes and form of glazing

A wide range of product samples and technical information is on display for reference purposes

## Role of the clerk of works

The clerk of works deals with

- visitors to the site
- drawings received
- deliveries
- instructions (see Clause 4 of the JCT Standard Form of Contract)
- details of any significant events, including any serious deficiencies in health or safety performance observed while on site.

The clerk of works liaises closely with contractor's staff. They must, however, maintain their independence, as they are responsible for working in the best interests of their employer or client.

## Skills and personal qualities

A clerk of works should:

- have a wide understanding of the building industry, including knowledge of materials, trades, methods and legal requirements
- be physically fit
- have a good head for heights
- be attentive to detail when checking work and materials
- be technically competent
- have good spoken and written communication skills
- be honest and vigilant, to make sure that the work and materials meet the required standard
- be able to establish an appropriate working relationship with the contractor's staff
- be persuasive and diplomatic, while remaining independent
- have good judgement, because they have to decide when to insist on corrections, when to persuade or negotiate and when to compromise
- set an example by acting in a professional manner at all times, including the wearing of personal protective equipment when on a construction site.

## Role of the clerk of works

The clerk or works should keep up to date with changes in construction methods and statutory legislation and carry out continuing professional development (CPD)

Member status of the Institute of Clerks of Works is open to those who have successfully achieved one of the following:

- NVQ/SVQ Site Inspection level 4
- A relevant BTEC/SQA higher national award
- A relevant first or higher degree
- Corporate membership by examination of one of the associated professional institutions recognised for exempting qualifications.

# Chapter 2

# Developing Construction Teams

## Contents

2.1  Team building                           25
2.2  Matrix organisation in practice         30

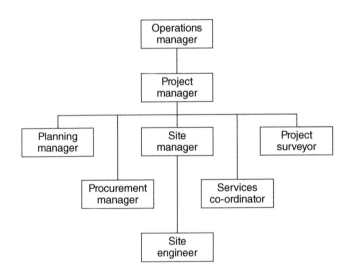

*Management of Construction Projects*, First Edition. Brian Cooke.
© 2015 John Wiley & Sons, Ltd. Published 2015 by John Wiley & Sons, Ltd.

## 2.1 Team building

Team building is an essential component of the site management process. Good project managers build competent teams of key personnel that they can carry from project to project. Creating harmony and developing a good working relationship with other team members is essential for the success of the project and the company alike.

A project team may be described as a number of people who work closely together to achieve a common goal (Burke).

The development of a number of teams to achieve specific objectives during a large project is essential for success of the project.

Examples are now shown of developing specific site teams:

- site management team
- procurement team
- surveying team
- site engineering team

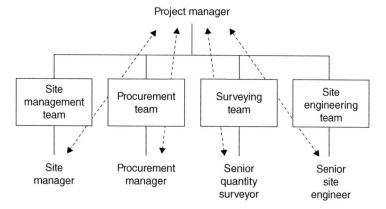

**Relationships between the project manager and the team leaders**

Team leaders are responsible for motivating their teams and reporting back to the project manager of any major problem areas.

### Developing a site management team

Key team members:

project manager
site managers
planning engineer
senior/site engineer
subcontractor representatives

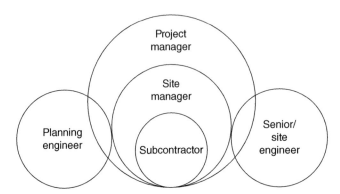

**Site management relationships**

Project manager – responsible for holding the team together and motivating the team by regular review and control meetings.

Site managers – key members of the site team. Responsible for implementing the short-term requirements of the programme, and arranging progress/review meetings with subcontractors in order to achieve weekly and monthly goals. Reports direct to the line manager.

Planning engineers – responsible for monitoring the programme both long term and short term. Directly involved in weekly progress reporting or updating long-term programmes and liaison with subcontractors and representatives. Reports on monthly progress situations and forecasts the effect of delays on the project completion date.

Site engineering – responsible for maintain control of line and level, working directly with site management personnel and subcontractors.

Subcontractors – maintain planned programme and progress goals. Attend weekly and monthly progress review meetings in order to maintain the programme.

## Developing a procurement team

Key team members:

project manager
procurement manager
planning engineer
contractors quantity surveyor
work package subcontractor

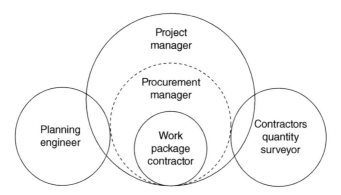

**Procurement team relationships**

Project manager – needs to be kept informed of the procurement situation regarding the progress with each of the work package contractors.

Contractors QS – responsible for administering work package contractors and agreeing the contract terms, variations, valuations and contractual matters.

Planning engineer – responsible for integrating the work package subcontractors into the programme, monitorz progress and reportz to the project manager at regular intervals.

Procurement manager – responsible for co-ordination of the work package contractors from tender to completion. Establishs design responsibilities with individual work package contractors. Agrees overall procurement programme with the planning engineer. Co-ordinates with the quantity surveyors regarding the contracts award and payment procedures.

Work package contractor – (the most important member of the team) unless they produce the work packages within the time, quantity and cost parameters, the team may fail to meet their objectives.

## Developing a surveying team (quantity surveyors)

Key team members:

commercial manager or senior surveyor
project manager
senior site quantity surveyor
assistant surveyors
subcontractors' representative surveyors

The site quantity surveying team falls under the control of a commercial manager based in the company's regional office. The project manager has direct access to the commercial manager regarding any contractual matters.

The project manager has direct control of the surveying personnel engaged on the project. The commercial manager may visit the site to assess the performance of his site-based surveying team.

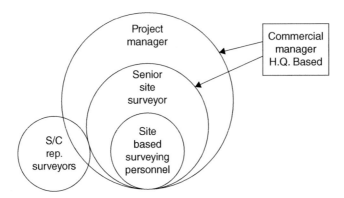

Commercial manager – responsible for all the surveying personnel within the company. Liaises with individual project managers regarding the surveying team engaged on the project.

Project manager – responsible for analysing monthly cost/value reports prepared by the senior site surveyor. Liaises with surveying team on contractual matters.

Senior site surveyor – manages surveying processes on site including:

- valuation process
- maintaining records of variations to contract
- liaison with professional quantity surveyor (client's representatives)
- payments to subcontractors and suppliers
- preparation of monthly cost/value reports
- reporting to the project manager

Assistant/trainee surveyors – undertake general site surveying tasks.Liaise with subcontractors representatives. Keeping site records as directed by the senior surveyor

Project surveying team – senior quantity surveyor, who is responsible to the commercial manager based in head office and to the project manager on site.

The senior surveyor is responsible for delegating responsibilities to the site surveying team, and on multi-million-pound management and design-and-build projects, there might be four or five site-based surveyors.

Towards the end of each month the senior surveyor will be responsible for preparing the cost/value reconciliation report (CVR). This involves reconciling the project value with the project costs to the date of the last valuation or month end.

Great skill – or luck – is required in forecasting the value to completion and the cost to completion as the contract nears its end. The senior surveyor assumes the role of a magician and often has to produce the proverbial 'white rabbit out of a hat' to save the contract from financial disaster.

## Developing a site engineering team

Key team members – this is clearly dependent on the size of the contract and the extent of the setting out work to be undertaken. Extensive use is now made of contract engineers, who are employed as "setting out machines". The larger projects usually employ a senior site engineer, or engineers, and engineering assistants.
 Site engineering team:

- project manager
- site manager
- senior engineer
- assistant engineers
- trainee engineers

Site engineering is often the initial training ground for site management personnel. Liaison with the site managers is essential in order to arrange for setting out work to be undertaken ahead of various trade gangs and subcontractors.

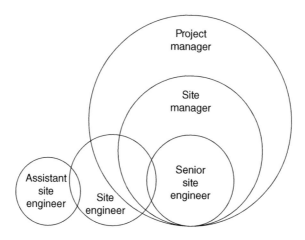

The site engineer takes instructions from the senior engineer in conjunction with the site manager.
 The assistant site engineers assist the site engineer or senior site engineer in daily setting out procedures. They work closely with the subcontractor's managers and foremen at the work face.

## 2.2   Matrix organisation in practice

The following note has been abstracted from the BAM Construction website. This relates to the company's approach to project organisation in implementating a matrix structure.

### Extract from reference to organisation structure

Lately, the company has adopted a matrix organization structure for the management of projects, which involves dynamic organization of project teams in a matrix structure. Responsibility for project implementation lies with teams as opposed to individuals. Each team comprises members drawn from both the technical department of the company as well as administrative ones. Further project teams consist of members drawn from third parties such as subcontractors, consultants and project managers. Operations of the project teams are guided by project charters evolved by requisite teams according to the development of each project. The project charter comprises:

- objectives of the teams
- project deliveries
- communication planning
- roles and responsibilities
- deadlines

A number of construction management textbooks mention "matrix management" and most fail to explain it clearly. Reading their explanations rarely advances an understanding of how it is applied in practice.

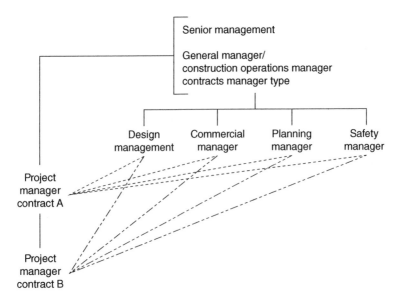

The project manager's project team includes:

* assistant project manager or site manager
* planning engineer
* senior quantity surveyor
* design team co-ordinator

The site project team can obtain advice from, or discuss problems with, their project manager, and also consult with their superiors at head office.

Many references to matrix management indicate the possibility of creating divided loyalties between members of the project team and their functional supervisors. This situation is often more evident with the senior site quantity surveyor who has a loyalty to both the commercial manager and the project manager.

A matrix diagram included in many references is illustrated here.

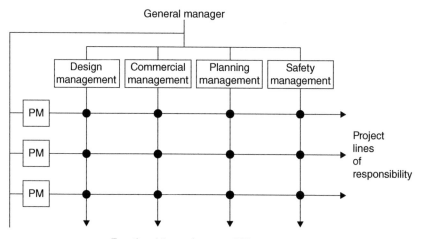

Interface between teams and the functional elements of an organisation

## Matrix management

**Definition** – A style of management where an individual has two reporting supervisors (bosses) one functional and one operational.

This is a common feature seen in project management, where an engineer, for example, reports to the chief engineer functionally and to the project manager on operational project issues.

Generally the reporting relationship is stronger because the functional manager controls the individual's evaluation.

For a matrix management style of organisation to be effective, the functional and operational managers must have equal weight in controlling the individual in their matrix.

www.strategicfutures.com/matrix management

This site indicates a network of interfaces – the matrix model is a network of interfaces between teams and the functional elements of an organisation. An interesting matrix management relationship diagram is shown between functional managers and the project manager.

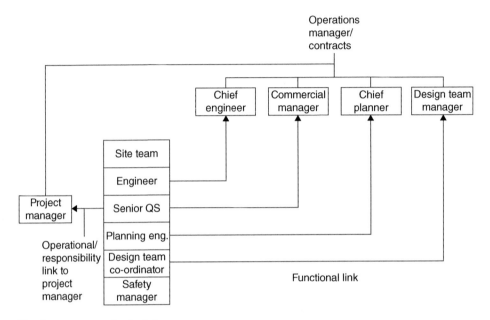

**Matrix management principles**

www.wikipedia.org

This excellent reference defines matrix management and illustrates the principles of a matrix organisation, and clear advantages and disadvantages are indicated.

www.global-integration.com/matrix-management-training

This reference defines a matrix organisation by saying that "a matrix structure simply describes a structure that helps the horizontal flow of information. Employers from different teams can be called upon for different projects according to the needs of the business."

## References

R. Burke – *Project Management, 4th Edition*, 2003, J. Wiley and Sons
Principles of developing a project team structure is outlined. The purpose of project team members and the benefits of developing teams is clearly outlined. A critical appraisal of "why teams win" and "why teams fail" is outlined.
R. Heller and T. Hindle – *Essential Manager's Manual*, 1998 *Edition*, Dorling Kindersley Publication

See the chapter on 'Managing Teams' – An excellent practical reference for the student and manager alike. It also has sections with first-class diagrams on areas covering:

- understanding how teams work
- setting up a team
- improving team efficiency
- working for the future.

All in all, a complete practical approach to team development.

Cole G.A. – *Management Theory and Practice, 6th Edition*, 2004, Thompson Publications.

References are made to Adair's work on effective team building (publ.1986).

Fryer B. – *The practice of construction management – People and building Performance, 4th Edition*, 2006, Blackwell Publications

See Chapter 7 – Group behaviour and teamwork. Contains excellent sections on group teamwork, performance and behaviour. Features of a good team, roles and leadership areas are clearly explained. This textbook contains a very comprehensive section on developing teams.

R. Burke – *Project Management. Planning and Control Techniques – 4th Edition* 2003. Wiley

An excellent practical explanation of matrix organisation structure, with clear bullet points on advantages and disadvantages

Cole G.A. – *Management Theory and Practice – 6th Edition*, 2012. DPP publications

*A Guide to the Project Management Body of Knowledge (PMBOK)*, Project Management Institute, ISBN 1-880410-23-0

# Chapter 3

# Project Planning – Linked Bar Charts and Procurement Programmes

## Contents

| | | |
|---|---|---|
| 3.1 | Overview of programming techniques | 35 |
| 3.2 | Procedure when using a linked bar chart programme | 36 |
| 3.3 | Procurement programmes | 39 |
| 3.4 | Procurement programme principles | 40 |
| 3.5 | Procurement symbols used on programmes | 42 |
| 3.6 | Extract from a procurement programme | 43 |

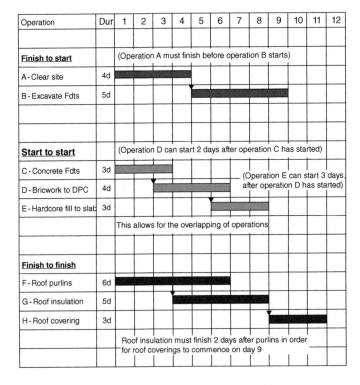

*Management of Construction Projects*, First Edition. Brian Cooke.
© 2015 John Wiley & Sons, Ltd. Published 2015 by John Wiley & Sons, Ltd.

## 3.1  Overview of programming techniques

The bar chart display was developed by Henry Gannt in 1915. Other systems, such as critical path analysis (CPA), precedence diagrams and line of balance came some forty years later.

The critical path method was developed by E.I. DuPont de Nemours & Co. in 1956. Further development work by Mouchley, Ketley and Walker in 1957 led to wider applications suitable for the industry. The onset of computers in the early 1960s led to analysis being undertaken by computers as a central analysis source.

The principles of network analysis are still included in construction management degree courses in order to develop an analytical approach to construction situations – long may this practice continue.

Critical path software developed by Pertmaster/Primavera is still used – although only applicable to the large projects involving complex relationships. There was no evidence of it being used on the projects included in the building case studies.

Precedence diagrams were developed to present a clear relationship between operations: finish to start, start to start and finish to finish. This enabled bar chart displays to more clearly illustrate these links in practice. As laptops became more powerful, speedy analysis on portable computers has aided the development of packages of the Powerproject type.

Linked bar charts are now the norm on construction projects. All the case studies reflect this, as they all use Asta Developments Powerproject and Team Plan software on their projects. The use of linked bar chart software has been encouraged by the training programmes available from Asta Developments. Any university offering construction and project management degree courses can obtain a free licence from Asta Developments for use within a university environment. The development of programming using linked bar chart software should now become the norm on university courses.

Construction firms tend to use trained planning engineers to undertake all aspects of planning during a project. They are responsible for preparing the programmes for pre-tender, pre-contract, master contract and procurement and also for programme updates and progress reports.

In some organisations the site manager is trained to use Powerproject, and often works with the project planner and takes responsibility for the short-term planning at site level.

Powerproject Version 12 is available from the Asta Development website. A computer download titled "Getting Started" is clearly presented and guides the user through the various stages of preparing a bar chart display.

Other linked bar chart software is available from Project Commander and others.

## 3.2  Procedure when using a linked bar chart programme

### Display 1

### Decision and tasks

Sort out:
Calendar display: days/weeks/months
Start date
Finish date
(any pre-contract period to be shown)Study project drawings

### Display 2

Establish WBS
(work breakdown structure)
Division of operations to be displayed:
Main work sections foundations etc.
Operations within work sections
Site strip
Excavate + concrete foundations
Assess operational durations in days/weeks etc.

### Display 3

Enter WBS on the bar chart
Main work section followed by operations within section
Enter operational durations (in days, weeks etc.)
See example display

### Display 3 (from Asta Display)

| Line | Operation | Dur | Start | Finish | June 1 | 2 | 3 | 4 | 5 | 6 | 7 | 8 | Month Wk No. |
|------|-----------|-----|-------|--------|--------|---|---|---|---|---|---|---|---------------|
| | Durations in weeks<br>Start date 1st June | | | Project - Title | | | | | | | | | 5 day week |
| 1 | FOUNDATIONS | | | | | | | | | | | | |
| 2 | Clear site | 2w | 1st June | 14th June | ███ | | | | | | | | |
| 3 | Exc/Conc Fdt | 1w | 1st June | 7th June | █ | | | | | | | | |
| 4 | Bkw DPC | 2w | 1st June | 14th June | ███ | | | | | | | | |
| 5 | H/C fill | 1w | 1st June | 7th June | █ | | | | | | | | |
| 6 | GF slab | 3w | 1st June | 21st June | █████ | | | | | | | | |
| 7 | SUPERSTRUCTURE | | | | | | | | | | | | |
| 8 | Bkw DPC GF-1st floor | 4w | 1st June | 28th June | ███████ | | | | | | | | |
| 9 | PC first floor | 2w | 1st June | 14th June | ███ | | | | | | | | |
| 10 | Bkw 1st-2nd floor | 4w | 1st June | 28th June | ███████ | | | | | | | | |
| 11 | PC 2nd floor | 2w | 1st June | 14th June | ███ | | | | | | | | |

## Work breakdown structure

Main work sections

Foundations – clear site and excavate and concrete foundations
　　　　　　　brickwork DPC / hard core fill
　　　　　　　ground floor slab
Superstructure – brickwork ground floor – 1st floor
　　　　　　　precast concrete first floor
　　　　　　　brickwork 1st floor – 2nd floor
　　　　　　　precast concrete second floor
This is a simplistic approach to work Breakdown Structure (WBS)

## Display 4 (linking the operations on the bar chart)

This is based on finish to start, start to finish and finish to finish relationships which are illustrated.

These relationships are illustrated in the project planning example displays using Power Project software by Asta development plc.

Asta developments supply the Power Project software package to consultants, contractors and individuals. An excellent back up service including in-company training facilities are available to users.

Reference: Asta Developments plc
Email: enquiries@astadev.com

## Display 4 (linking on bar chart)

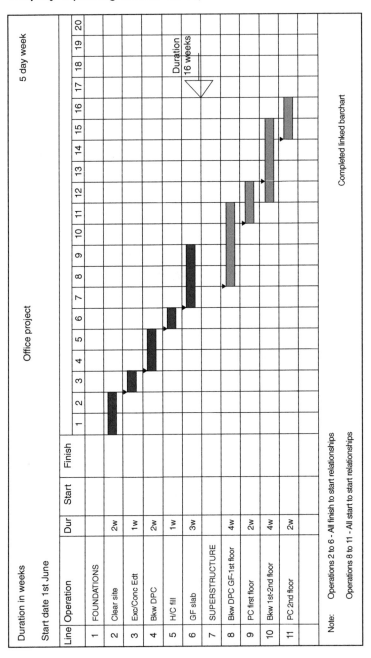

Duration in weeks

Start date 1st June

Office project

5 day week

| Line | Operation | Dur | Start | Finish |
|------|-----------|-----|-------|--------|
| 1 | FOUNDATIONS | | | |
| 2 | Clear site | 2w | | |
| 3 | Exc/Conc Edt | 1w | | |
| 4 | Bkw DPC | 2w | | |
| 5 | H/C fill | 1w | | |
| 6 | GF slab | 3w | | |
| 7 | SUPERSTRUCTURE | | | |
| 8 | Bkw DPC GF-1st floor | 4w | | |
| 9 | PC first floor | 2w | | |
| 10 | Bkw 1st-2nd floor | 4w | | |
| 11 | PC 2nd floor | 2w | | |

Duration
16 weeks

Note: Operations 2 to 6 - All finish to start relationships

Operations 8 to 11 - All start to start relationships

Completed linked barchart

## Relationships when linking operations

| Operation | Dur | 1 | 2 | 3 | 4 | 5 | 6 | 7 | 8 | 9 | 10 | 11 | 12 |
|---|---|---|---|---|---|---|---|---|---|---|---|---|---|
| | | | | | | | | | | | | | |
| **Finish to start** | | (Operation A must finish before operation B starts) | | | | | | | | | | | |
| A - Clear site | 4d | | | | | | | | | | | | |
| B - Excavate Fdts | 5d | | | | | | | | | | | | |
| | | | | | | | | | | | | | |
| | | | | | | | | | | | | | |
| **Start to start** | | (Operation D can start 2 days after operation C has started) | | | | | | | | | | | |
| C - Concrete Fdts | 3d | | | | | | | | | | | | |
| D - Bricwork to DPC | 4d | | | | | | (Operation E can start 3 days after operation D has started) | | | | | | |
| E - Hardcore fill to slab | 3d | | | | | | | | | | | | |
| | | This allows for the overlapping of operations | | | | | | | | | | | |
| | | | | | | | | | | | | | |
| **Finish to finish** | | | | | | | | | | | | | |
| F - Roof purlins | 6d | | | | | | | | | | | | |
| G - Roof insulation | 5d | | | | | | | | | | | | |
| H - Roof covering | 3d | | | | | | | | | | | | |
| | | Roof insulation must finish 2 days after purlins in order for roof coverings to commence on day 9 | | | | | | | | | | | |

These relationships are practical links between related operations occurring during the construction process.

Start to start relationships allow the overlapping of related operations. Overlaps are difficult to express on a network diagram as activities need splitting into separate parts. This complicates the network (circle and link) diagram and may prove complex to analyse.

## 3.3 Procurement programmes

The preparation of procurement programmes is an integral part of the programming and control process. This is of particular importance on design and build and management contracts.

In many contract situations the main contactor is responsible for appointing, organising and managing named and domestic subcontractors.

Key work packages need careful assessment of "lead-in times" prior to commencing work on site. The lead-in time is the time period between the completion of the procurement process and the commencement of work on site.

The site construction programme and the procurement programme are illustrated for two steelwork operations. Links are shown between the master programme and the procurement programme. Displaying the procurement programme above the master programme is a useful presentation technique adopted by a number of main contractors.

## 3.4 Procurement programme principles

Before starting the steelwork erection on site there are three stages in the procurement cycle:

- obtaining a competitive quotation for undertaking the work
- allowing for the steelwork design period
- allowing time for manufacture of the steelwork prior to delivery to site

A two-week lead-in time is shown on the display for steelwork operations on the office block, and three weeks on the main portal frame.

Milestone symbols may be integrated into the display to indicate:

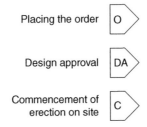

**A range of procurement milestone stages are illustrated, together with various hatching formats which may be introduced on bar line displays.**

## Procurement programme principles

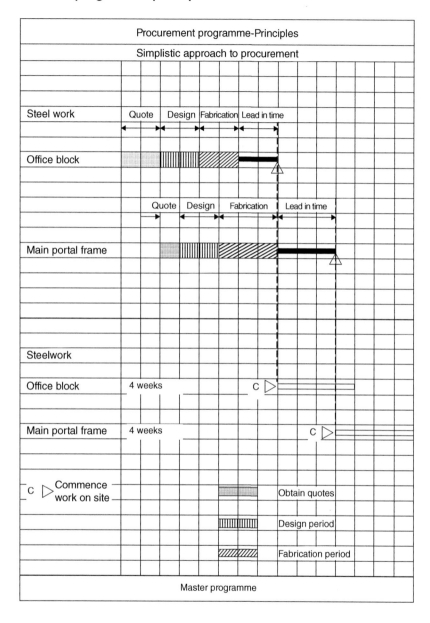

## 3.5 Procurement symbols used on programmes

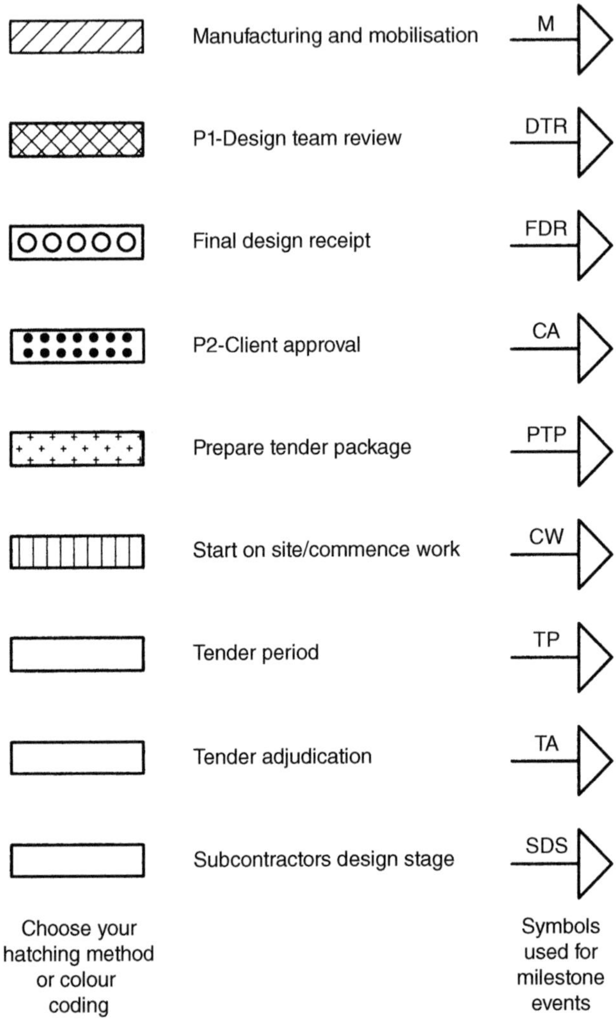

| | | | |
|---|---|---|---|
| ///// | Manufacturing and mobilisation | M ▷ | |
| ⊠⊠⊠⊠ | P1-Design team review | DTR ▷ | |
| ○○○○○ | Final design receipt | FDR ▷ | |
| ●●●●●●● | P2-Client approval | CA ▷ | |
| +·+·+·+ | Prepare tender package | PTP ▷ | |
| ▥▥▥ | Start on site/commence work | CW ▷ | |
| ☐ | Tender period | TP ▷ | |
| ☐ | Tender adjudication | TA ▷ | |
| ☐ | Subcontractors design stage | SDS ▷ | |

Choose your
hatching method
or colour
coding

Symbols
used for
milestone
events

## 3.6   Extract from a procurement programme

(based on Chinley School project)

An extract from the procurement programme is illustrated. Operation 8 from the programme has been presented in bar chart format for the six procurement stages. The procurement displays have been developed using Powerproject and /Team Plan commercial software.

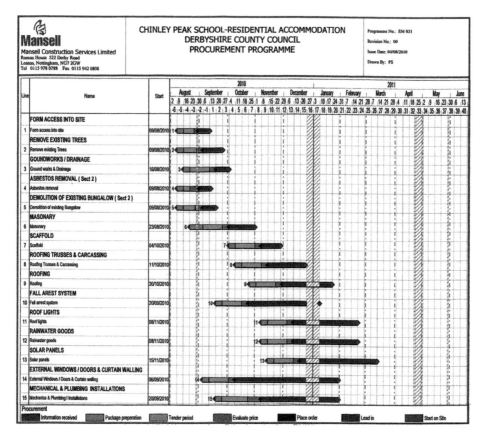

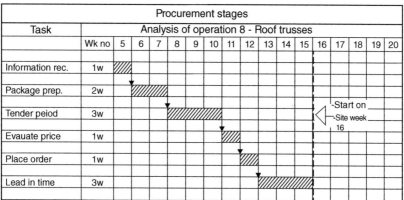

# Chapter 4

# Contracts

## Contents

| | | |
|---|---|---:|
| 4.1 | Schedule of current JCT standard contracts – 2011 | 45 |
| 4.2 | JCT standard building contract – 2011 | 45 |
| 4.3 | JCT major projects construction contract – 2011 | 47 |
| 4.4 | JCT design and build contract – 2011 | 48 |
| 4.5 | JCT management building contract – 2011 | 51 |
| 4.6 | JCT construction management contracts – 2011 | 53 |

*Management of Construction Projects*, First Edition. Brian Cooke.
© 2015 John Wiley & Sons, Ltd. Published 2015 by John Wiley & Sons, Ltd.

## 4.1 Schedule of current JCT standard contracts – 2011

The following summary has been summarised from the JCT Contracts website. Relationships between the parties involved in the contract are illustrated to aid student understanding.

Ref: www.jctltd.co.uk/category/contract-families

Twelve contract forms are indicated with main features outlined relative to projects using selected contracts.

The contracts included in the JCT contract families include:

- Standard building contract – 2011
- Intermediate building contract – 2011
- Minor works building contract – 2011
- Major projects construction contract – 2011
- Design and build contract – 2011
- Management building contract – 2011
- Construction management contract – 2011

JCT – CE contract – 2011
Measured term contract – 2011
Prime cost building contract – 2011
Repair and maintenance contract – 2011
Home owner contracts – 2011

The seven contact case studies indicated in the text have been undertaken using the contracts indicated:

2 projects – Design and Build (Single stage tender)
3 projects – Design and Build (Two stage tender)
1 project – Construction management contract
1 project – JCT minor works

Short summary notes on the six highlighted contracts will be developed (*).

## 4.2 JCT standard building contract – 2011

Designed for large or complex projects where detailed contract provisions are needed. Suitable for projects procured via competitive tenders or negotiated contracts.

### Features

- Employer responsible for the design. – Supplied by the clients architect or design team working on the employers behalf.
- Standard building contracts have an optional provision for a "contractors design portion", if the contractor is to be responsible for the design of a specific part of the works.
- Depending on the type of standard form the employer will need to provide drawings / specification / work schedules or bills of quantities. This enables the client to specify the quantity and quality of work at the tender stage.
- The employer may use the standard form with a quantities provision or the contractor may be required to assess his own quantities based on the information shown on the drawings.
- Standard building contracts are administered by the architect, quantity surveyor or a contract administrator (the butcher / baker / candlestick maker scenario).

## JCT Intermediate building contract – 2011

Designed for projects involving all recognised trades where fairly detailed contract provisions are needed but without complex building services or other specialist work.

### Features

- If the appointed contractor is to be responsible for designing specific parts of the works then an intermediate building contract with contractor's design must be used.
- Employer will need to provide drawings and bills of quantities as before. When using the intermediate building contract with contractors design, the employer must also detail the requirements for the parts of the works that the contractor is responsible for designing.
- Contract again administered by the architect / quantity surveyor or a contract administrator.

The minor works building contract is used where the work is of a simple nature. It is not suitable where the project is complex enough to require bills of quantities or detailed control procedures.

## Relationships between parties – standard building contract

```
                          ┌──────────┐
                          │  Client  │
                          └──────────┘
                               │
                               │          Client's representative
                               ├───────── architect/quantity surveyor
                               │          or contract administrator
                               │
             ┌─────────────────┴─────────────────┐
      ┌─────────────┐                    ┌────────────────────┐
      │ Design team │                    │ Building contractor │
      └─────────────┘                    └────────────────────┘
             │                                     │
         Architect                        Project manager/
             │                              site manager
             │                                     │
      ┌──────┴──────┐                      ┌───────┴───────┐
  Structural     Quantity             Domestic          Named
  engineer       surveyor           subcontractors   subcontractors
             │
        Services
        engineer
```

Client to provide: - drawings, specifications
work schedules
bills of quantities

Contractor: - Contractor's designed portion
for specific parts of the works
if applicable

**JCT Standard Form of Contract Relationships**

Ross and Williams outline in a simplistic form, the types and financial implications of the various standard forms of building contracts.

These include lump sum contracts, re-measurement contracts (measure and value arrangements) and cost reimbursement contracts. Practical applications are outlined in relation to examples from building practice (reference to Chapter 4 – contracts and documentation.)

## 4.3   JCT major projects construction contract – 2011

Designed for large scale construction projects where major works are involved (no minimum or maximum contract value ranges are indicated).

Used by employers who regularly procure large scale construction work. The work is carried out by contractors with experience and ability to take greater risk than would arise under other JCT contracts.

Major project construction contracts are suitable for the design and build method.

### Features

- Contractor responsible for the design as well as completing the works.
- The contract may be based on **a concept provided by the employers advisors**. Alternatively the contractor may be responsible for providing and completing the design from the outset (either through the contractors in-house team or engaging design specialists.)
- Often a "novation agreement" is put in place so that the architect or designer who initially worked on the scheme with the employer continues to complete the design under the responsibility of the main contractor.

### Authors note

This contract follows the concept of the design and build arrangement, applied to the larger type project.

This may be based on an "Employer led design and build arrangement" or a "Contractor led design and build relationship".

Separate relationship diagrams are illustrated for both these situations in the notes on Design and Build.

## 4.4   JCT design and build contract – 2011

The design and build contract is designed for construction projects where the contractor carries out both the design and construction work.

- The scale of design work needed to be carried out by the contractor can vary greatly on design and build projects.

Sometimes the contractor will complete the design based on the concept provided by the employer's advisors. Novation of the clients design team is common practice and this arrangement can be expressed as "Client Led Design and Build".

Other arrangements may call for the contractor to be responsible for producing and completing the design right from the outset. The arrangement is often expressed as "Contractor Led Design and Build". Refer to relationship diagrams for the above situations.

- Adequate time is needed to allow detailing of the employers requirements to ensure that design requirements are fully met.
- The employer will normally use a clients representative to administer the contract (project manager etc.). Both the client and contractor follow procedures contained in the PIBA Plan of Work in relation to the stages in the design and construction process.

A number of bar chart displays based on the Plan of Work will be included to clarify the sequence of events during the design / tender and construction process. These bar chart displays will be included in a separate section.

## Client led / design and build relationships

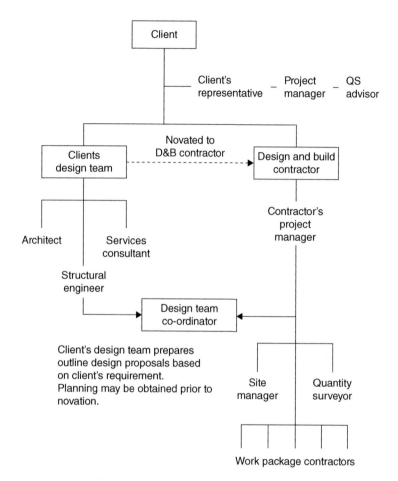

**Design and build relationships**

## Contractor led relationships

The Design Team Co-ordinator is an essential member of the contractors management team.

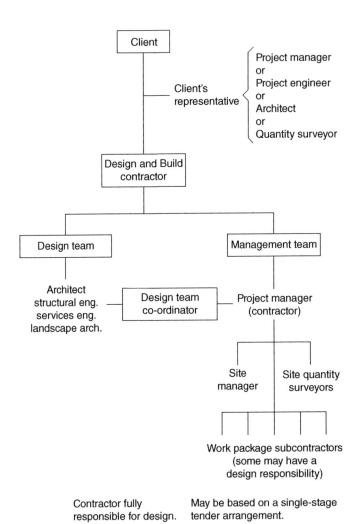

Contractor fully responsible for design.

May be based on a single-stage tender arrangement.

**Design and build relationships**

## 4.5 JCT management building contract – 2011

This contract is designed for construction projects where the employer appoints a management contractor to oversee the works.

Construction is completed under a series of separate works contractors, which the management contractor appoints and manages for a fee.

### Authors note

A wide range of major projects have been completed in the north west region using the management contract these include:

Manchester Airport – Terminal 2
Manchester Airport – Terminal 3
Manchester NEM Arena – Concert venue
Liverpool One (parts) – Office and apartment blocks
Trafford Centre – Shopping complex

### Features

- Suitable for large scale complex projects where flexibility and an early start is required on site. i.e. package subcontractors such as earthworks, piling and basement work may commence as soon as the subcontract packages are awarded. Work contracts for the frame, floors and cladding can commence as the project progresses.
- Work on management building contracts are carried out in packages and design often runs in parallel to site activity. This allows for greater flexibility and more control for employer over design.
- The management contractor employs the work package contractors to carry out the construction work. The subcontractors are directly accountable to the management contractor.
- Contracts are established between the management contractor and each work package subcontractor. The management contractor pays the work package subcontractors on a monthly bases.

- An extensive team of design team co-ordinators and surveyors are necessary to keep a check on information flow and ensure prompt monthly payments to the subcontractors. Accounts are finalised as each work package is completed.
- The client often engages an individual project manager to act on his behalf as the clients representative. Alternatively a company specialising in managing large projects may be engaged as the clients advisors.
- The employer is responsible for the design and this is normally supplied to the management contractor by the architect or design team working on the employers behalf. Hence the need for a design team co-ordinator to form part of the management contractors management team.
- The management contractor agrees monthly valuations with each of the work package subcontractors and the client pays them through the management contractor (as the management contractor has a contact with each of the subcontractors).

A relationship diagram showing the links between the various parties is illustrated.

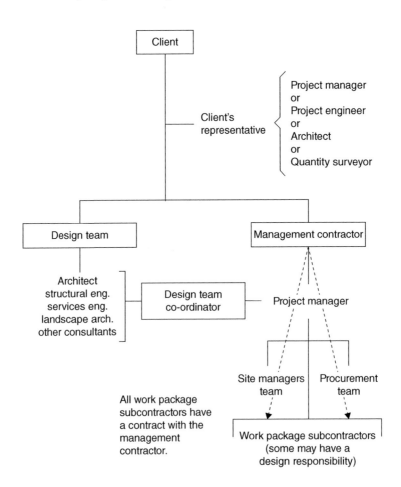

A key to the success of a management contract is the programming / planning procedures developed to control work package subcontractors.

## 4.6  JCT construction management contracts – 2011

This is for use where the employer appoints separate trade contracts to carry out the works and a separate construction manager to oversee the completion of works for a fee.

### Features

- The construction management contract occupies a suite of documents to cover agreement between the construction manager (CM/A) and the construction management trade contractors (CM/TC).
- The contract used where a separate responsibility for design and construction of the project is desired.
- The employer produces the design and enters into separate trade contracts with the suppliers (or subcontractors) to carry out the construction work.

The employer is responsible for paying the separate trade contractors on a monthly basis.

- The construction manager (an individual or an organisation providing a project management service) is appointed to manage the project and act as an agent on the employers behalf. He is responsible for issuing instructions, making decisions and preparing certificates (for each of the trade contractors).
- The construction manager administers the conditions of the trade contracts.

Projects undertaken on a management contract basis in the north west:

Manchester City Football Stadium
(Client Manchester city council – Management Contractor – Laing O'Rouke)

## JCT management contract – 2011

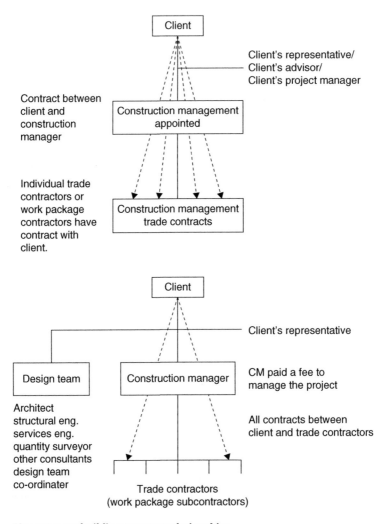

**Management building contract relationships**

## Reference

Ross and William (2013) Financial Management in Construction, Wiley Blackwell.

# Chapter 5
# Materials Management

**Contents**

5.1 Site planning to reduce loss and waste     56
5.2 Bad site practice observations     56
5.3 Good site practice for materials management     64

*Management of Construction Projects*, First Edition. Brian Cooke.
© 2015 John Wiley & Sons, Ltd. Published 2015 by John Wiley & Sons, Ltd.

## 5.1 Site planning to reduce loss and waste

A good site manager, given the resources, can ensure successful materials storage and handling. But a good site manager without the resources can achieve very little.

Materials management on an open spacious site is far easier than on a confined site. Management attitudes to the control of materials varies widely from one company to another. Often managers on the smaller projects receive little encouragement from senior managers who are more concerned about profitability than reducing waste on site.

It is essential to allocate specific monies in the tender for the costs of managing materials – no mater which of the preliminaries it is hidden in or included in. Monies need to be included in a tender bid to provide:

- hard standing areas for site access and materials storage areas
- facilities to cover and protect components from the weather
- timber pallets or crates in which to store materials such as formwork, scaffold fittings and drainage fittings
- a storage compound with dedicated material storage areas for the storage of window frames, lintels, pipes and sheeting etc.
- lifters or loaders to handle materials onto scaffolds and loading platforms
- tarpaulin sheets to cover plasterboard and other sheet material
- racked storage for specialist timber products, trunking and pipework.

The list is endless and often specific to the building programme.

Care must be taken to ensure that over-ordering of materials does not occur. This applies to materials such as faced or rubbed stonework, insulation materials, plasterboard linings and cladding trims.

A variety of images are shown of materials remaining on site due to over-ordering situations – if only the management were aware of these situations!

Waste management systems should be set up that match quantity ordered against quantity used and bill quantities, and an attempt be made to flag up the differences. This is of particular importance for ready-mixed concrete items in order to monitor additional quantities used due to the increased thickness of foundations and such like.

## 5.2 Bad site practice observations

Here are a series of site images indicating both good and bad site practice in relation to materials management. The project case studies all include aspects of materials control and site layout planning situations. Every site situation is different and there isn't one magical solution to them all.

As the site progresses from foundation work to roofing and external works, different materials are brought on site for each construction stage. The site layout plan should therefore be flexible enough to accommodate this ever-changing situation.

Here first is a summary of the various snapshots from miscellaneous sites.

### Snapshot one – high-quality housing project

There is no excuse for the degree of untidiness on the site. Areas under scaffolds should be maintained clear of waste material as each scaffold platform is raised. On speculative housing projects extensive use is made of labour-only subcontractors. The main contractor should ensure that they are responsible for the material losses that they create. A "name-and-shame" approach should be taken, and they should be contra-charged for cleaning up their mess at the end of each stage of work.

### Snapshot two – surplus materials and over-ordering

The over-ordering of materials during a contract generates extensive costs for the contractor. The images display surplus bricks, plastic piping, insulation and timber within the compound area. Packs of sheet insulation, expensive earthenware pipes and aluminium or galvanised metal external cladding materials await disposal. Not only has the contractor been charged for delivering the goods to site – but he now has to pay for them to be disposed of.

### Snapshot three – materials management of foundation blockwork

The problem has arisen due to lack of site preparation prior to delivery of the foundation blockwork. The area around the foundation should have been stoned up or protected with Teram sheeting on which to position the blocks.

Stacked blocks have also been positioned too close to the top of the trench, causing the blocks to fall into the excavation. The management cannot blame the occurrence on weather conditions – even though they may try to.

### Snapshot four – storage of blockwork

Prior to the delivery of major materials, storage areas should be prepared to protect the bottom courses of bricks and blocks. Materials should not be stored on sloping ground which is liable to destabilise the block stacks. Blocks can be left on timber pallets on level ground.

Block stacks should be unbanded with metal cutters in preference to being hit with a shovel to break the bands.

### Snapshot five – materials management on the smaller project

The images illustrated can readily be viewed from footpaths adjacent to the sites. The projects were being undertaken by small contractors with evidently very little concern regarding materials control on the projects. The projects are possibly using labour-only subcontractors with the site management unaware of the material wastage they were creating.

A name-and-shame approach would simply have no effect on these contractors. Perhaps they are making too much profit to be concerned about materials control.

## Materials management is all about...

**"The Good"**

**"The Bad"**

**"The Ugly"**

## Snapshot one – high quality housing project

**Materials mismanagement on a high quality residential project. Surely the subcontractors cannot be held responsible for all this mess!**

## Snapshot two – surplus materials and over-ordering

**A disorganised storage compound – situations of this type occur towards the completion of a project, often resulting from the over-ordering of materials**

**Surplus materials lying around the site at the contract completion – surpluses again created by over-ordering of materials during the contract**
**Readers are advised to assess the cost implications of the four images shown.**

## Snapshot three – materials management of foundation blockwork

An example of loss of material control when storing blockwork for use in shallow foundations.

Typical mismanagement when site management staff are pressurised

**A total lack of site preparation prior to the delivery of materials. Blocks unloaded and placed on sides of excavation – no consideration of the protection or preparation prior to delivery. Trench sides collapse due to increased loading.**

**Ask the question – who is responsible?**

## Snapshot four – storage of blockwork

Problem areas with five speculative houses to be constructed with labour-only sub-contract gangs.

**Complete lack of site preparation prior to the delivery of site materials – blocks delivered and placed on uneven ground (on pallets)**

**Block stacks inadequately banded – as the bands are removed the blocks are allowed to fall over**

**The subcontractor cannot be held fully responsible as a prepared area for storage of materials was not provided by the speculative contractor.**

## Snapshot five – materials management on the smaller project

Materials control is often ignored on the smaller projects. Management "close their eyes" to the losses occurring on site, but they must be aware of the untidiness of their sites. Even a passer-by can clearly see the chaos from the footpath adjacent to the sites.

Images from the footpaths

**Mixed materials stockpile – packaging, pallets, timber off-cuts, plasterboard, insulation – why not put it in a skip?**

**Timber off-cuts from roof purlins and rafters – mass waste of treated timber (this could have been easily deposited in large bags for disposal)**

**Mixed blocks, London bricks, demolished materials, timber and ready-mixed concrete**

## 5.3 Good site practice for materials management

*Construction Practice* (by the same author) was published in 2011. This book has an extensive range of images illustrating both good and bad practice of material management.

Good practice observations are indicated here, together with a comprehensive section on the use of waste skips to separate the collection of waste materials on site.

On larger projects, specialist waste disposal firms are often engaged to manage waste on site. Monthly waste analysis reports are prepared on the amount of waste disposed of each month. The aim is to achieve a maximum of 5% waste to be returned to landfill.

An example of the waste services provided by Premier Waste (a north-west company) is illustrated in *Construction Practice*.

### Materials management practices illustrated in the case studies

Each case study illustrates the proposed and actual site layouts. The use of skips to ensure the separation of waste on a site is a common practice on major projects. Many companies appreciate the benefits of cleaning and rebanding materials such as formwork for reuse on the next project.

The responsibility for waste management is often that of the assistant site manager. He is responsible for ensuring that the waste management policy is implemented and that subcontractors and labour-only gangs are responsible for the waste they create.

A name-and-shame whiteboard, prominently displayed, highlights subcontractors failing to respond to site waste procedures. This is considered a convenient way of making them comply to site standards of waste management.

## Materials management – good practice examples

Use of storage container bins – easy to stack and handle containers of scaffold parts

Well-planned storage areas prior to delivery – storage on pre-prepared areas – blocks and water stillage

Scaffold and formwork fittings, stored in containers

Wall panel formwork, stacked clear of ground after use

Stackable, see-through containers for small fittings

Oil spillage platform for the location of fuel drums, avoids fuel spillage problems

## Materials management

Cleared area in front of concrete frame used to recover formwork material and scaffolding

**Materials used for formwork support cleared and banded for transfer to another site**

**Scaffolding and formwork support system racked up for removal from site**

# Materials management – materials storage and handling

Dedicated storage areas are used for scaffolding sub-contractors, materials are stored in open metal boxes and fittings stored in stackable crates. Scaffold planks and tube are stored adjacent to erection areas (on a timber-framed project).

**Recycling timber**

**Formwork timbers being recovered for reuse on a further project. Materials cleaned and branded for ease of handling**

## Materials management – materials storage and handling

**Use of waste containers for handling mixed materials – in this case a mortar skip**

**Large plastic storage bags used for collecting materials – to be handled by fork lift trucks**

**Various types of metal containers and skips in use on a housing project**

## Materials management – good storage practice

**Allowances at tender stage provide adequate monies to provide good site compound facilities**

**Taylor-Wimpey site showing good site materials control policy**

## Organisation of projects in Australia – November 2012

## Observation of construction projects in Australia

Projects were well organised from a site layout point of view. Sites were tidy with planned access facilities and good access roads. Extensive use was made of coloured edge barriers to identify access routes. The separation of work areas, material storage areas and site compounds was clearly defined.

Compliance with PPE requirements appeared to be rigorously enforced

Coloured road barriers were also used for marking material storage areas in specified site locations. It is interesting to note the use of precast concrete boxes located alongside the site access road.

**Coloured road barriers used to identify access roads**

**Road barriers set up to contain specific sub-contractors' materials and equipment.**

**Materials stored on prepared ground adjacent to access roads**

**Extensive use of projecting storage skips (conti boxes) on the facade of a building – the conti boxes can be used for storage of components positioned by tower crane or other mobile cranes**

Sites were well planned and material management was considered a key to their success.

# Chapter 6

# Mechanical Handling and Risk Assessment

## Contents

| | | |
|---|---|---|
| 6.1 | Crane selection factors | 74 |
| 6.2 | Crane types – mobile and tower cranes | 75 |
| 6.3 | The telehandler (telescopic handler) | 79 |
| 6.4 | The aerial work platform (scissor lift) | 81 |
| 6.5 | Hydraulic lifter – cherry picker | 84 |
| 6.6 | Schedule of risk assessment areas | 87 |
| 6.7 | Plant utilisation for major stages of work | 91 |
| 6.8 | Student task – method statement and risk assessment | 93 |

*Management of Construction Projects*, First Edition. Brian Cooke.
© 2015 John Wiley & Sons, Ltd. Published 2015 by John Wiley & Sons, Ltd.

## 6.1 Crane selection factors

**Crane selection**

The selection of lifting equipment is influenced by:

- site access conditions and space around the building (especially boundary space situations)
- form of construction, i.e. steel frames, in-situ concrete or precast frame
- area or radius to be covered by crane
- height of building
- weight of critical loads at maximum radius of crane
- ground conditions and type of base required for crane
- lifting restrictions imposed by site location – access for delivery vehicles
- proximity hazards – overhead power lines or moving loads over adjacent buildings or roads.

**Inclusion of plant allowances at tender stage**

Plant and equipment is included in this estimate build-up on a time and related fixed cost assessment. A breakdown of the costs is required for valuation purposes.

A major item of plant such as a tower crane would include fixed costs such as:

- delivery to site
- provision of crane base
- erection of crane
- provision of power and services
- dismantling of crane.
- removal from site

and time-related costs:

- hire of crane and drivers
- fuel and servicing.

Monies for plant are claimed at each monthly valuation stage. The contractor is required to provide the client's surveyors with a breakdown of plant costs for valuation purposes. The majority of plant costs are included in the contract preliminaries section of the bill.

**Alternative plant and method proposals – value engineering**

On competitive tenders the contractor may propose monetary savings resulting from an analysis of alternative methods of construction. This may bring savings in plant costs. Value engineering tasks are undertaken in order to provide a more competitive tender at the final tender stage. Analysis of the proposed construction methods may result in a reduction in the final tender sum.

Refer to the value engineering example in relation to the office and hotel project in Section 8.8.

Plant items and risk assessments include:

- crane types: mobile cranes and tower cranes; needs a risk assessment for heavy plant and machinery
- telehandler
- aerial work platform (scissors lift), plus risk assessments
- hydraulic lifter (cherry picker), plus risk assessments
- schedule of risk assessments including:
  access and egress
  delivery of materials
  personal protective equipment
  storage of materials
- plant utilisation for major stages of work.

## 6.2   Crane types – mobile and tower cranes

Mobile cranes may be tracked (similar to those on an excavator) or lorry mounted.

**Large tracked mobile crane with fly jib between two buildings**        **Lorry mounted crane with hydraulic jib**

**Large tracked crane with lattice jib used for heavy lifting operations**

**Track mounted mobile crane fitted with an extended fly jib**

**Luffing jib tower crane (to prevent crane moving over adjacent properties and roads etc.)**

**Self-erecting tower crane with horizontal jib (horizontal jib tower cranes are also available)**

## Hydraulic mast tower crane

**A self erecting tower crane with a telescopic tower (luffing jib tower crane for use on a restricted access site)**

## Risk assessment – heavy plant & machinery (mobile and tower cranes)

| Work Activity | Heavy Plant & Machinery | | Risk Assessment No. | | 11 |
|---|---|---|---|---|---|

Identify persons or maximum number of people exposed together with frequency / duration of task

- All persons
- Every day
- Various hours

Hazards identified (potential to cause harm to people)

| | |
|---|---|
| 1 | Chippings from break up - eyes |
| 2 | Flying particles |
| 3 | Noise |
| 4 | Being struck by plant / machinery |
| 5 | Unauthorised access and use |
| 6 | Dust /fumes |
| 7 | Unstable ground |
| 8 | Adverse weather |
| 9 | Crushing / entanglement |
| 10 | Overhead / underground obstructions |

### Initial Risk before Control Measures

| | A) Severity | | | B) Probability | | Risk (AxB)* |
|---|---|---|---|---|---|---|
| 1 | Negligible | | 1 | Unlikely | | |
| 2 | Slight | | 2 | Possible | | |
| 3 | Moderate | | 3 | Quite Possible | | 20 |
| 4 | Severe | | 4 | Likely | x | |
| 5 | Very Severe | x | 5 | Very Likely | | |

### Control Measures

| | |
|---|---|
| 1 | Trained certified drivers / operators |
| 2 | Visual checks of condition before use |
| 3 | Recorded weekly maintenance checks |
| 4 | All plant / machinery stored in suitable area to prevent unauthorised access and use |
| 5 | Suitable PPE to be worn at all times |
| 6 | Barriers & signage used used to create segregation between plant / mavhinery and people |
| 7 | Ensure plant / machinery is suitable for the task |
| 8 | Goalposts' used to highlight overhead obstructions e.g. cables, wires etc |
| 9 | Plant without an enclosed cab for the driver will not operate in adverse weather |

### Residual Risk after Control Measures

| | A) Severity | | | B) Probability | | Risk (AxB)* |
|---|---|---|---|---|---|---|
| 1 | Negligible | | 1 | Unlikely | | |
| 2 | Slight | | 2 | Possible | x | |
| 3 | Moderate | | 3 | Quite Possible | | 10 |
| 4 | Severe | | 4 | Likely | | |
| 5 | Very Severe | x | 5 | Very Likely | | |

Key to Risk Totals
0-8
9-15 Medium, no action required
16-25 High Risk

This refers to all types of heavy plant, including tracked excavators, lorry mounted or tracked cranes.

- Traffic routes for crane movement around the site should be separated by interlocking barriers.
- The hard standing for crane locations should be clearly identified on site plans.
- The site hazard board should highlight when the cranes are in use.

Dedicated unloading areas should be highlighted on the plans.

## 6.3 The telehandler (telescopic handler)

The telehandler is widely used in the construction industry. It is similar in appearance to a forklift truck. The telescopic boom can extend to a height of 9 metres.

A wide variety of attachments may be fitted to the end of the boom such as pallet forks, muck grab or lift table.

The pallet fork attachment can be used to move loads to and from places. Loads may be placed on loading platforms or roofs, which would otherwise need a crane.

Some machines are fitted with front stabilisers to avoid problems of tilting or becoming unstable when handling heavy loads.

Hazards are similar to those resulting from the use of other hydraulic lifters such as cherry pickers and hydraulic platforms.

A wide range of machines are available for hire

**Used for moving materials around the site**

**Large telehandler with hydraulic boom fitted with pallet forks**

## 6.4   The aerial work platform (scissor lift)

The hydraulic aerial work platform may be used for:

* handling and fixing curtain wall framing and glazed infill panels to a building facade
* fixing purlins to the roof of a steel portal frame
* handling a variety of materials – as used in filling gabion mesh cages etc.
* temporary access for people or equipment to inaccessible areas, usually at height
* work platform usually fitted with electrical outlets or a compressed air supply
* access for cleaning of building facades.

**Aerial work platforms being used for a wide range of tasks**

## Handling and fixing of curtain wall support and glazed panels to a five-storey building using an aerial work platform

Use of aerial work platform (hydraulic scissors lift) capable of lifting to 5 or 10 storey

Curtain wall support can be raised and fixed from hydraulic platforms. The platform is large enough to accommodate glazed panels and fixing gang.

This is one of the most efficient methods of handling curtain wall support. On the larger platforms stillage stands may be positioned on the platform for supporting glazed panels.

## Risk assessment – mobile elevated work platform (scissor lift)

| Work Activity | Mobile Elevated Work Platform | | Risk Assessment No. | | 15 |
|---|---|---|---|---|---|
| Identify persons or maximum number of people exposed together with frequency / duration of task | | | | | |
| • | All persons | | | | |
| • | Every day | | | | |
| • | Various hours | | | | |
| Hazards identified (potential to cause harm to people) | | | | | |
| 1 | Falls from platform | | | | |
| 2 | Fall of materials | | | | |
| 3 | Collision with structure, persons or other machinery | | | | |
| 4 | Collapse of platform | | | | |
| 5 | Mechanical failure of platform | | | | |
| 6 | Unauthorised access / use | | | | |
| 7 | Adverse weather | | | | |

| Initial Risk before Control Measures | | | | | | |
|---|---|---|---|---|---|---|
| A) Severity | | | B) Probability | | | Risk (AxB)* |
| 1 | Negligible | | 1 | Unlikely | | |
| 2 | Slight | | 2 | Possible | | |
| 3 | Moderate | | 3 | Quite Possible | | 20 |
| 4 | Severe | | 4 | Likely | x | |
| 5 | Very Severe | x | 5 | Very Likely | | |

| Control Measures | |
|---|---|
| 1 | Ensure work area is cordoned off to prevent persons walking around the area |
| 2 | Only trained operatives to use platform |
| 3 | When moving platform, the area must be kept clear and movement controlled by a trained operator |
| 4 | Platform regularly maintained and inspected (LOLER) |
| 5 | Prior to use, inspection regime to be completed and recorded |
| 6 | Platform not to be used in adverse weather conditions |
| 7 | Platform stored and kept secure to prevent unauthorised access / use |

| Residual Risk after Control Measures | | | | | | |
|---|---|---|---|---|---|---|
| A) Severity | | | B) Probability | | | Risk (AxB)* |
| 1 | Negligible | | 1 | Unlikely | | |
| 2 | Slight | | 2 | Possible | x | |
| 3 | Moderate | | 3 | Quite Possible | | 10 |
| 4 | Severe | | 4 | Likely | | |
| 5 | Very Severe | x | 5 | Very Likely | | |

| Key to Risk Totals |
|---|
| 0-8 |
| 9-15 Medium, no action required |
| 16-25 High Risk |

- Hard standing areas must be provided around working areas.
- Area of work must be cordoned off to prevent operative access.
- Scissors platform must not be overloaded.

Applications – fixing glazing panels, curtain walling support systems, transporting and raising materials in skips (see gabion wall images in chapter 9, in the Merlin Project case study).

## 6.5 Hydraulic lifter – cherry picker

**Short-arm lifter**

**Long-arm lifter**

**Track mounted lifter with stabilisers**

**Long-arm lifter**

**Track mounted lifter with stabilisers**

Used for difficult access situations.

Extensively used in the erection of steelwork, as illustrated in the following images.

## Hydraulic lifters – cherry pickers in action

**Three hydraulic lifters in a site compound**

**Erecting steelwork to a six-storey frame**

**Fitting bolted connections from platform**

## Risk assessment – hydraulic lift boom (cherry picker)

| Work Activity | Hydraulic Lift Boom (Cherry Picker) | | Risk Assessment No. | | 16 |
|---|---|---|---|---|---|

**Identify persons or maximum number of people exposed together with frequency / duration of task**

- All persons
- Every day
- Various hours

**Hazards identified (potential to cause harm to people)**

| | |
|---|---|
| 1 | Falls from platform |
| 2 | Fall of materials |
| 3 | Collision with structure, persons or other machinery |
| 4 | Collapse of platform |
| 5 | Mechanical failure of platform |
| 6 | Unauthorised access / use |
| 7 | Adverse weather |

**Initial Risk before Control Measures**

| | A) Severity | | | B) Probability | | Risk (AxB)* |
|---|---|---|---|---|---|---|
| 1 | Negligible | | 1 | Unlikely | | |
| 2 | Slight | | 2 | Possible | | |
| 3 | Moderate | | 3 | Quite Possible | | 20 |
| 4 | Severe | | 4 | Likely | x | |
| 5 | Very Severe | x | 5 | Very Likely | | |

**Control Measures**

| | |
|---|---|
| 1 | Ensure work area is cordoned off to prevent persons walking around the area |
| 2 | Only trained operatives to use platform |
| 3 | When moving platform, the area must be kept clear and movement controlled by a trained operator |
| 4 | Platform regularly maintained and inspected (LOLER) |
| 5 | Prior to use, inspection regime to be completed and recorded |
| 6 | Platform not to be used in adverse weather conditions |
| 7 | Platform stored and kept secure to prevent unauthorised access / use |
| 8 | All operatives to wear a safety harness |

**Residual Risk after Control Measures**

| | A) Severity | | | B) Probability | | Risk (AxB)* |
|---|---|---|---|---|---|---|
| 1 | Negligible | | 1 | Unlikely | | |
| 2 | Slight | | 2 | Possible | x | |
| 3 | Moderate | | 3 | Quite Possible | | 10 |
| 4 | Severe | | 4 | Likely | | |
| 5 | Very Severe | x | 5 | Very Likely | | |

**Key to Risk Totals**
0-8
9-15 Medium, no action required
16-25 High Risk

### Application areas

All types of hydraulic lifters are available, capable of lifting to a height of 6–8 metres. There are short-arm and long-arm lifters operated from hydraulic platform by operator, and they are extensively used for handling glazed wall panels and fixing brackets at the corners of units.

Extensive space is required around the base of the machine for storing materials, and good access is necessary around work areas.

## 6.6 Schedule of risk assessment areas

| Ref No. | Activity | Tick Areas |
|---|---|---|
| 1 | Access and Egress | |
| 2 | Breaking Concrete | |
| 3 | Collapse of Scaffold | |
| 4 | Compressor and Breaker | |
| 5 | Delivery of Materials | |
| 6 | Diesel | |
| 7 | Engine Oil | |
| 8 | Excavations | |
| 9 | Exposure to Asbestos | |
| 10 | Handling of Concrete | |
| 11 | Heavy Plant & Machinery | |
| 12 | Hot Working Inc. Welding & Burning | |
| 13 | Hydraulic Oil | |
| 14 | Manual Handling | |
| 15 | Mobile Elevated Work Platform | |
| 16 | Hydraulic Lift Boom (Cherry Picker) | |
| 17 | Noise | |
| 18 | Other Site Personnel | |
| 19 | Paving | |
| 20 | Personal Protective Equipment | |
| 21 | Storage of Materials on Site | |
| 22 | The Use of Vibrating tools | |
| 23 | Use of Disc Cutters and Abrasive Wheels | |
| 24 | Welding | |
| 25 | Use of Hand Tools | |
| 26 | Use of Step Ladders | |
| 27 | Use of Mobile Scaffold Towers | |
| 28 | Use of Portable Electrical Equipment | |
| 29 | Vehicle Movement on and of site | |
| 30 | Welfare Facilities | |
| 31 | Working from Heights | |
| 32 | Use of Lifting Genie | |
| 33 | Use of Hammer Drill | |
| 34 | Exposure to HAVS (hand arm vibration syndrome) | |
| 35 | Counterbalanced Fork Lift Trucks / telehandler | |
| 36 | Fork Reach Trucks | |
| 37 | Employee Health Surveillance | |
| 38 | | |
| 39 | | |
| 40 | | |

The following plant activities relate to most risk assessments:

- access and egress
- delivery of materials
- personal protective equipment
- storage of materials

This may prove of interest as it lists 37 risk assessment areas.

## Risk assessment – access & egress

| Work Activity | Access & Egress | | | Risk Assessment No. | | 1 |
|---|---|---|---|---|---|---|
| Identify persons or maximum number of people exposed together with frequency / duration of task | | | | | | |

- All persons
- Every day
- Various hours

| Hazards identified (potential to cause harm to people) |
|---|
| 1 | Falling Objects |
| 2 | Dust |
| 3 | Noise |
| 4 | Slips,trips and falls |
| 5 | Movement of heavy plant |
| 6 | Adverse weather conditions |

**Initial Risk before Control Measures**

| A) Severity | | | B) Probability | | | Risk (AxB)* |
|---|---|---|---|---|---|---|
| 1 | Negligible | | 1 | Unlikely | | |
| 2 | Slight | | 2 | Possible | | |
| 3 | Moderate | | 3 | Quite Possible | | 16 |
| 4 | Severe | x | 4 | Likely | x | |
| 5 | Very Severe | | 5 | Very Likely | | |

**Control Measures**

| 1 | All vehicles are to be marshaled when entering or leaving the site |
|---|---|
| 2 | The frequency of vehicles entering the site shall be strictly controlled |
| 3 | Site access and egress arrangements will be included in Site Safety induction |
| 4 | There will be suitable fencing / protective barrier installed |
| 5 | Traffic Management System will be in place |
| 6 | Suitable signage wll be displayed in prominent positions |

**Residual Risk after Control Measures**

| A) Severity | | | B) Probability | | | Risk (AxB)* |
|---|---|---|---|---|---|---|
| 1 | Negligible | | 1 | Unlikely | | |
| 2 | Slight | | 2 | Possible | x | |
| 3 | Moderate | | 3 | Quite Possible | | 8 |
| 4 | Severe | x | 4 | Likely | | |
| 5 | Very Severe | | 5 | Very Likely | | |

**Key to Risk Totals**
0-8
9-15 Medium, no action required
16-25 High Risk

Access and egress is an important risk assessment item. A number of contractors arrange for a dedicated site area for the access of plant where steelwork and utilities are to be lifted and stored.

The site traffic management plan indicates access roads, hard standings, unloading areas and pedestrian assess routes.

The daily hazard board will highlight restricted access areas during specific site operations.

# Risk assessment – delivery of materials

| Work Activity | Delivery of Materials | | | Risk Assessment No. | | | 5 |
|---|---|---|---|---|---|---|---|

**Identify persons or maximum number of people exposed together with frequency / duration of task**

- • All persons
- • Every day
- • Various hours

**Hazards identified (potential to cause harm to people)**

1. Moving vehicles
2. Damage to existing property
3. Slips,trips and falls
4. Material handling
5. Unsafe load
6. Access and egress
7. Cuts and abrasions

### Initial Risk before Control Measures

| | A) Severity | | | B) Probability | | Risk (AxB)* |
|---|---|---|---|---|---|---|
| 1 | Negligible | | 1 | Unlikely | | |
| 2 | Slight | | 2 | Possible | | |
| 3 | Moderate | | 3 | Quite Possible | x | 12 |
| 4 | Severe | x | 4 | Likely | | |
| 5 | Very Severe | | 5 | Very Likely | | |

### Control Measures

1. Trained / certified
2. Maintenance check of the vehicle
3. Marshalled vehicles entering and leaving ths site
4. Barriers and suitable signage in place
5. Deliveries made at agreed times
6. Designated storage set up prior to deliveries

### Residual Risk after Control Measures

| | A) Severity | | | B) Probability | | Risk (AxB)* |
|---|---|---|---|---|---|---|
| 1 | Negligible | | 1 | Unlikely | | |
| 2 | Slight | | 2 | Possible | x | |
| 3 | Moderate | x | 3 | Quite Possible | | 6 |
| 4 | Severe | | 4 | Likely | | |
| 5 | Very Severe | | 5 | Very Likely | | |

**Key to Risk Totals**
0-8
9-15 Medium, no action required
16-25 High Risk

Areas for materials storage should be marked on the site layout plan. Vehicle drivers unloading materials should be familiar with the location of storage areas.

## Risk assessment – personal protective equipment

| Work Activity | Personal Protective Equipment | | | Risk Assessment No. | | 20 |
|---|---|---|---|---|---|---|

Identify persons or maximum number of people exposed together with frequency / duration of task

- All persons
- Every day
- Various hours

Hazards identified (potential to cause harm to people)

| | |
|---|---|
| 1 | Falling objects |
| 2 | Slips, trips and falls |
| 3 | Moving site vehicles |
| 4 | Cuts and abrasions |
| 5 | Nuisance dust |
| 6 | Chemicals / solvents |

### Initial Risk before Control Measures

| | A) Severity | | | B) Probability | | Risk (AxB)* |
|---|---|---|---|---|---|---|
| 1 | Negligible | | 1 | Unlikely | | |
| 2 | Slight | | 2 | Possible | | |
| 3 | Moderate | | 3 | Quite Possible | | 20 |
| 4 | Severe | x | 4 | Likely | | |
| 5 | Very Severe | | 5 | Very Likely | x | |

### Control Measures

| | |
|---|---|
| 1 | Compulsory wearing of helmets, safety boots and hi-visibility clothing |
| 2 | Wearing all other PPE as and when required |
| 3 | Training in the use of PPE |
| 4 | Storage of PPE in company vehicles |
| 5 | Specific PPE for high risk work (e.g. harness when working at height) |

### Residual Risk after Control Measures

| | A) Severity | | | B) Probability | | Risk (AxB)* |
|---|---|---|---|---|---|---|
| 1 | Negligible | | 1 | Unlikely | | |
| 2 | Slight | | 2 | Possible | x | |
| 3 | Moderate | | 3 | Quite Possible | | 8 |
| 4 | Severe | x | 4 | Likely | | |
| 5 | Very Severe | | 5 | Very Likely | | |

**Key to Risk Totals**
0-8
9-15 Medium, no action required
16-25 High Risk

It is a requirement on all construction sites that site visitors must wear full personnel protective equipment – safety helmet, safety eye protection, gloves, safety footwear and a yellow vest top. They must also be accompanied around the site by a member of the site team.

## 6.7 Plant utilisation for major stages of work

**Liverpool Man Island project**

This is a 15-storey steel frame with two sliding staircase cores.

In-situ concrete sliding formwork was used for the staircase cores, with the horizontal jib tower crane used for concreting the top five slide pours and for general material handling.

Steelwork erection to the skeleton frame was undertaken with telescopic mobile cranes, and for the steelwork connections, long-armed hydraulic lifters (cherry pickers) were used.

**General view of staircase towers and steel frame**

**Telescopic mobile crane and hydraulic lifters(Genia Z type) in action, Erecting the steel frame around the staircase cores**

## Stockport College project

This has a four-storey steel frame with adjacent single-storey workshop unit.

A rubber-tyred mobile crane with telescopic jib worked in conjunction with hydraulic lifters.

Castilated beams were used to support metal deck floors.

**Four-storey steel frame**

**Single-storey frame erected with separate hydraulic mobile crane and lifters**

**Single-storey castilated roof beams – erected by mobile crane and hydraulic lifters (cherry pickers)**

## 6.8   Student task – method statement and risk assessment

a. Prepare a method statement and risk assessment for the steel structure to be erected at the third-floor level.
   State all assumptions to clarify your approach.
b. Prepare a risk assessment for the operations involved.

Access is available from the adjacent building at third floor level. Space is available in front of the building for unloading and the storage of steelwork – access is available to all building elevations.

**Side elevation**

**Front elevation**

## Student task

**View of completed fourth floor steelwork**

# Chapter 7

# Managing Construction Defects

## Contents

| | | |
|---|---|---|
| 7.1 | Managing defects | 96 |
| 7.2 | Recording defects during a project | 96 |
| 7.3 | Dealing with defects at the handover stage of a project | 103 |
| 7.4 | Dealing with defects at the end of a defects liability period | 104 |

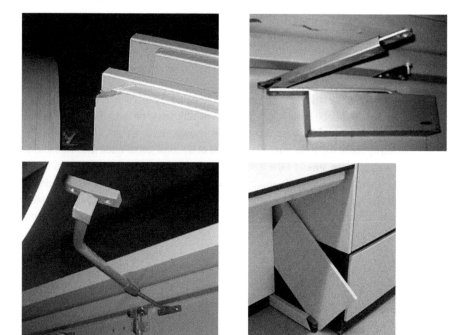

*Management of Construction Projects*, First Edition. Brian Cooke.
© 2015 John Wiley & Sons, Ltd. Published 2015 by John Wiley & Sons, Ltd.

## 7.1 Managing defects

Managing defects involves carrying out planned inspections at each stage of the construction process. For example, during the fitting out of kitchens and bathrooms regular inspections should be recorded at each stage of work: first fix, second fix and final fix. This process should reduce the items contained on the final snagging list.

The approach to recording and managing defects during a project is to be covered under the following headings.

* recording defects during the project – a photo image approach
* dealing with defects at handover stage of a project – practical completion
* dealing with defects at the final completion stage

The problems associated with collateral warranties will also be addressed.

## 7.2 Recording defects during a project

The use of photo images as evidence of building defects is an effective approach to defects management during a project. The defects record sheet indicates:

* a description and photo image (ensure that the photo image clearly indicates the problem and is close enough to show the detail clearly)
* responsibility for the defect: design/detail fault, subcontractor/workmanship problem or damage by others
* comments on specific responsibility regarding carrying out remedial works.

Where the cause of the defect cannot be linked directly to a workmanship problem or to a specific subcontractor, the contractor will ultimately be required to bear the cost. For example, on the defects record sheet shown, occurrence 02 could have been caused by some mysterious passing vehicle or by some other trade moving materials.

The contractor should take a systematic approach to the inspection procedure because it may be difficult to access items which are later covered by other trades – for example, service runs that are later inaccessible after plasterboarding and plastering.

Contractors tend to use their own record sheets for recording and monitoring defects.

SHEET No.10

PROJECT – CENTRAL FLATS

DATE – 10 Feb 2012

| Ref | Description | Report Date | Detail of defect (photo image) | Responsibility | | | Comments |
|-----|-------------|-------------|-------------------------------|----------------|--|--|----------|
| | | | | Design | Subcontractor | Others | |
| 01 | Damaged corners to fibreboard ceiling panels | 10/2 | 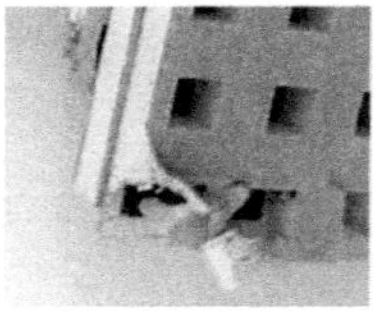 | | Masons N/SC | | Damaged fibreboard panels to be replaced at S/C expense |
| 02 | Damaged angle bead to side of door return | 10/2 | 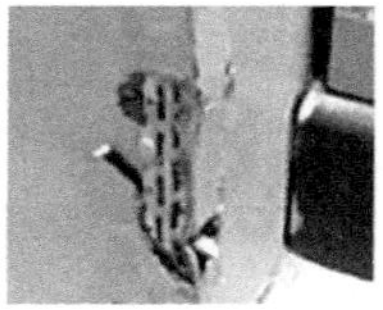 | | E&R Plaster D/S/C | | Subcontractor to replace damaged angle bead. Protection to plaster returns requires attention |
| 03 | Holes in ceiling around electrical fittings | 10/2 |  | | Hart Conc. Co. Contractor Chainage | | Holes in concrete floor wrongly positioned by concrete frame subcontractor |

**Defects record sheet – photo image approach**

**EXAMPLES OF DEFECTS – involving snagging, incomplete work and defects.**

## Recording defects – 1

The wrong location of service holes for pipes, lighting points, ventilation ducts and such like may become the responsibility of the client's design team.

Defects record sheets may be prepared by a site manager or assistant manager as part of their routine observations. A systematic approach to recording the observations in a specific defects file is essential.

The recording of photo images is standard practice with a growing band of clerks of works.

Typical headings from a clerk of works record sheet include:

- references
- issue/defect identified
- date reported
- photo image display
- responsibility: design or contractor
- risk identification
- further measures
- date defect was completed

A similar photo image approach can be applied to defect checks during the installation or fixing of an item while work is in progress.

Similar procedures may be adopted at the snagging (handover) stage of the project. Where on these occasions the clerk of works may be treated as "enemy number one". Please note, "The inspector may only be carrying out his job".

The management of defects that arise during a project is often difficult for the contractor to handle. It is therefore important to adopt a more organised approach to the recording of defects.

A defective work notice form is illustrated, as used by a large contracting organisation; each major company tends to develop their own approach to notifying action on defective work.

Where subcontractors fail to undertake the work, procedures are often in place to contra charge the subcontractor. This may sound a sensible approach in practice but recovering the costs once the subcontractor has left the site may prove difficult.

## Recording defects – 2

| INSPECTION SHEET | | Chinley School Project | | |
|---|---|---|---|---|
| SNAGGING | INCOMPLETE WORK | DEFECT | | |
| Date: | Room/Area | | | |
| No. | Location | Description | Works | |
| | | | Trade | Complete |
| | | | | |
| | | | | |
| | | | | |
| | | | | |
| | | | | |
| SNAGGING | INCOMPLETE WORK | DEFECT | | |
| | | | | |
| | | | | |
| | | | | |
| | | | | |
| | | | | |
| SNAGGING | INCOMPLETE WORK | DEFECT | | |
| | | | | |
| | | | | |
| | | | | |
| | | | | |
| | | | | |
| | | | | |

**A typical format for recording inspections between the main contractor and main subcontractors.**

This "inspection sheet" acts as a record of incomplete work/defects and snagging items.

## Recording defects – 3

| INSPECTION SHEET | | Chinley School Project | | |
|---|---|---|---|---|
| **SNAGGING** | INCOMPLETE WORK | DEFECT | | |
| Date: | Room/Area | | | |
| No. | Location | Description | Works Trade | Complete |
| 1 | Block 1 - R10 | Clean windows to rooms GR1 to GR7 | | |
| 2 | GF doors D1-6 | Adjust door latches to D1 to D6 Grd. floor | | |
| 3 | GF doors D1-6 | Ease and fix intumescent strips to doors D1 to D6 | | |
| 4 | R24/26/28 | Make good ceiling plaster around fire alarms | | |
| | | | | |

| SNAGGING | **INCOMPLETE WORK** | DEFECT | | |
|---|---|---|---|---|
| 1 | Ext. brickwork | Complete sealant fixing to expansion joint in external brickwork – corner of Block1 | | |
| 2 | Block 1 | Fix brackets to fire hydrant points (6 No.) | | |
| 3 | Block 1 | Complete internal painting around Velux roof lights | | |
| 4 | R16-21 | Complete grouting to floor tiling – Rooms 16–21 | | |

| SNAGGING | INCOMPLETE WORK | **DEFECT** | | |
|---|---|---|---|---|
| 1 | Staff area | Make good skirting board joints at wall returns in dining room – 1st foor staff area | | |
| 2 | GF floor | Daido rail out of horiz. align. in main lounge area | | |
| 3 | 1st floor corridor | Replace damaged door plates 8 doors | | |
| 4 | 1st floor storage cupboards | Replaster to sloping ceiling areas in four storage cupboards SC1, 6 & 8 | | |

## Examples of defects – snagging, incomplete work or defect

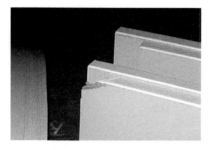

Defect item: corner of insulated feature panel damaged

Defect item: broken door closure – replace complete fitting

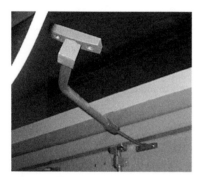

Snagging item: door stay requires refixing to frame

Snagging item: wrong position of service holes in floor – sleeve and make good around opening

Snagging item: make good around connections, prior to tiling

Incomplete work: refix panel at side of unit

## Defective work notice

---

### Defective Works Notice

REF: _____

Name of Contract: _____

Contract No: _____     Date:_____

Contractor: _____

Attention: _____

Address _____

_____

Dear Sirs,
You are notified that the works detailed below are not in accordance with your
contract requirements.
You must submit within _____ days, detailed proposals to remedy defects / correct
the defective work to the satisfaction of the contractor.
Should you fail to comply with this notice then we reserve the right to have yhe
necessary works carried out on your behalf and charge you in full costs incured.

| **Details of the Defective Work** |
| --- |
| |

Signed on behalf of: _____ Position: _____

| **Proposed Action to Correct** |
| --- |
| |

Signed on behalf                         Position:_____
of Contractor: _____

| **Remedial Works Agreed with Appropriate 3rd Party** |
| --- |
| See Correspondence Attached: |

Distribution: _____

## 7.3 Dealing with defects at the handover stage of a project

Before outlining the snagging and defects procedures at the practical completion stage of a project, an understanding of contract procedures is necessary. The relationship between the practical completion and final completion stages of a contract is illustrated diagrammatically. This indicates the defects snagging list prior to practical completion and the final defects schedule at the end of the defects liability period.

The defects situation at practical completion is not clearly defined in the contract. Practical completion is normally understood to be "when the works are complete for all practical purposes". Any outstanding items of work are considered as being only of a minor nature and would not clearly affect the proper functioning of the building.

Minor items of work may then be completed in the first few days after the handing over of the building to the client.

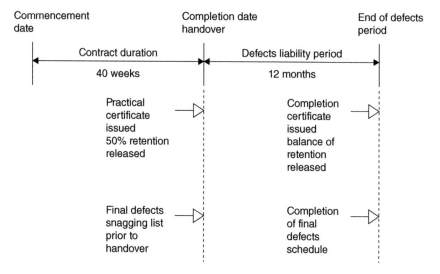

| Commencement date | | Completion date handover | | End of defects period |
|---|---|---|---|---|

Contract duration
40 weeks

Defects liability period
12 months

Practical certificate issued 50% retention released

Completion certificate issued balance of retention released

Final defects snagging list prior to handover

Completion of final defects schedule

**Contractual relationships**

At the handover stage, a certificate of practical completion will be issued by the architect. This will indicate the date of practical completion, and the quantity surveyor will release one half of the retention fund at the next payment stage. The defects liability period will then commence at the practical completion date.

An example defects inspection sheet is illustrated, which includes snagging items, incomplete work and defects to be identified during the inspection stages.

Good relationships are essential between the main contractor, clerk of works and the various subcontractors in order that defects identified are speedily dealt with.

## 7.4 Dealing with defects at the end of a defects liability period

The defects liability period (DLP), which is normally 12 months, is stated in the contract appendix. As the DLP comes to a close, a final snagging list will be prepared by the client's representatives.

Outstanding defects will need to be completed by the date specified, in order that a certificate of completion can be issued and the final retention released.

### Practical note

In some situations the main contractor or subcontractor may fail to return to complete the outstanding items at the end of the DLP. Some small subcontractors consider that it is not financially viable to complete outstanding items. In these cases the contractor would complete the items and deduct the cost from the retention due to the subcontractor (perhaps by this time the subcontractor may have gone out of business anyway).

Clauses are included in the contract which allow the client to "set off" monies against such occurrences. The client may therefore bring in "A.N. other" to complete outstanding items and deduct the monies from the contractor's final account.

# Chapter 8

# Hotel and Office Project Development

## Contents

| | | |
|---|---|---|
| 8.1 | Contents of the case study | 106 |
| 8.2 | Project overview | 107 |
| 8.3 | Project information | 109 |
| 8.4 | Client/contractor relationships | 110 |
| 8.5 | Site plan | 111 |
| 8.6 | Main contractors site organisation | 112 |
| 8.7 | Tender negotiation stage with preferred bidder | 112 |
| 8.8 | Value engineering proposals | 114 |
| 8.9 | Site layout planning | 118 |
| 8.10 | Hotel block – sequence of work for floors 5 to 7 | 126 |
| 8.11 | Programme for hotel building | 129 |
| 8.12 | Site monitoring of external brickwork operations | 133 |
| 8.13 | Office block – programme of work exercise | 136 |
| 8.14 | Student programming task | 139 |
| 8.15 | Completed office building | 141 |

*Management of Construction Projects*, First Edition. Brian Cooke.
© 2015 John Wiley & Sons, Ltd. Published 2015 by John Wiley & Sons, Ltd.

**Elevation of the completed building**

## 8.1  Contents of the case study

This inner city development of £12 million involves the construction of a five storey steel-framed office block alongside a six to ten storey precast crosswall hotel block.
  The case study covers the following aspects of the project.

- A description of the works for the office block and hotel building.
- Client/contractor relationships, together with the main contractor's site organisation structure.
- Value engineering proposals of the "preferred bid" contractor for this single-stage design and build project are outlined.
- The factors influencing crane selection are outlined as site space restrictions limited their choice and movement.
- Site layout planning, plus the approach to management of materials on site construction sequence for the erection of precast concrete crosswall frame for the hotel block.
- The programme for the office block sequence of work.
- Progress monitoring of the brickwork to the external elevations.
- The method of constructing the external brickwork outer wall, in relation to the value engineering proposals.
- The sequence of construction for the six-storey steel-framed office block.

Students and readers are then required to prepare a programme for the frame.
  Comments are also included on the value engineering proposals with respect to the external brickwork.

## 8.2   Project overview

The development consists of a five to eight storey precast concrete crosswall framed building to form a 150 bed hotel, and adjacent landscaped area and car parking. Adjacent to the hotel, a six storey steel-framed office block is to be

**Office building**

**Hotel building**

**View of office block with cladding to rear elevation in progress**

constructed and retail units are to be provided to the ground floor area of the office development.

The six storey office block incorporates retail units on part of the ground floor area. A plant room is to be constructed on part of the roof area. External claddings to three elevations of the building is to be formed in glass curtain walling, tied to the frame at each floor level.

The site compound and storage area is to be located on the new car park area and adjacent land is available for future developments.

The hotel block is of precast crosswall construction. The inner crosswalls and flank walls support precast widespan floor units. The loading from the precast structure is supported on a steel transfer floor located at the first floor level. The loads are then transferred down steel columns onto pad foundations. The ground floor building area is to contain a series of retail units on completion.

The foundation of the hotel buildings are to be constructed on CFA bored piles, pile caps, ground beams and suspended in-situ ground floor slab. The office block foundations are to be constructed on stabilised ground and simple pad foundations.

**Crosswall construction at third floor level**

**Steel transfer floor at first floor level**

## 8.3   Project information

Client: *Ibis Hotels – Royal London Asset Management Company*
Location: *Central Manchester*

Contract value: *£11.7 million*
Contract period: *56 weeks*
Commencement date: *June 2011*
Completion date: *July 2012*
Form of contract: *Design and build, single stage tender, novated design team*
Main contractor: *Galliford Try (Construction Northern)*

## 8.4 Client/contractor relationships

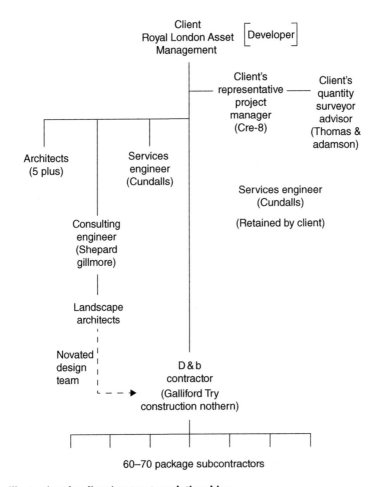

**Diagram illustrating the client/contractor relationships**

## 8.5 Site plan

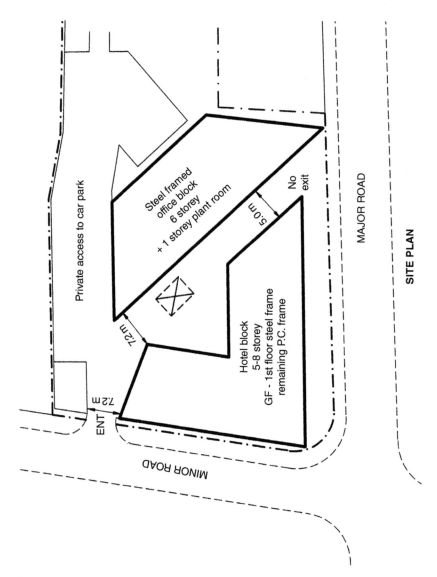

**Location of the hotel and the office block, indicating the dimensional restrictions between buildings for crane movement.**

## 8.6 Main contractors site organisation

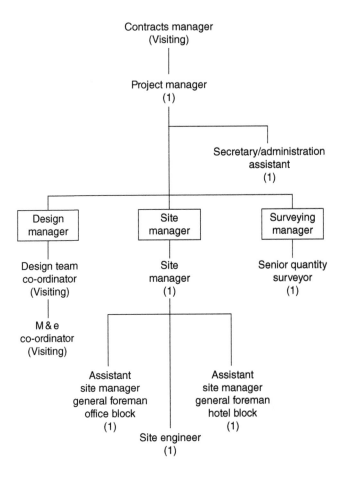

Diagram illustrating the main contractors site organisation

## 8.7 Tender negotiation stage with preferred bidder

Six major contractors were invited to submit competitive quotes based on the design and build contract (single stage tender arrangement). Galliford Try (Northern) were selected as the preferred bidder subject to the negotiation of the final tender sum.

The contractor's appointed project manager had been part of the tender team and directly involved in all the tender negotiations. He was therefore appointed as the "bid manager" to lead the value engineering proposals and present the final bid proposals to the client's team.

The contractor value engineered the tender to incorporate the following proposals:

- Proposal 1 – Revised foundation type to eliminate the CFA piling on the office block.
- Proposal 2 – Considerations of the steelwork design for the office block
- and transfer floor.
- Proposal 3 – Alternative system of external wall cladding.
- Proposal 4 – Consideration of alternative crainage methods for the erection of the steel and precast frames.

## Proposal 1 – alternative solution to the foundations to the office block

The original proposals specified the installation of in-situ concrete CFA bored piles with pile caps and ground beams supporting a 300 mm thick in-situ ground floor slab for both the office block and the hotel building.

Alternative proposals involved stabilising the ground to the office block area using the Bullivant NRG dynamic compaction process. This would enable simple pad foundations to be designed to support the office structure. The ground floor slab would correspondingly be reduced in thickness (from 300 mm to 150 mm) thereby offering further savings.)

## Proposal 2 – redesign the steelwork of the office block and the transfer floor of the hotel block

A redesign of the steelwork resulted in savings over the original design.

## Proposal 3 – alternative solution to the external envelope cladding

Original proposals were to provide a composite concrete wall panel to the external elevations. The concrete facade panel was to incorporate blue brick facings, cast on at the manufacturing stage. This would eliminate the need for an external scaffold provision.

## 8.8 Value engineering proposals

**Example of composite wall panels on a similar hotel project, with composite blue brick external storey-height panels**

**Example of an external skin of blue brick supported on metal angles fixed to the precast flank walls**

The contractor's proposals involved providing a 150 mm precast external wall panel to the precast frame. The blue facing brick exterior wall would be supported on stainless steel angles fixed to the outside face of the precast wall panels at each floor level. This proposal would need an external scaffold.

A cost study indicated that this proposal would be considerably cheaper than the composite panel faced in blue brickwork (see the images of proposed support system on the following pages).

## Proposal 4 – consideration of alternative crainage methods.

The tender was based on providing two lorry mounted hydraulic jib cranes: Crane 1 would be used to erect the 5/6 story office, and Crane 2 would be simultaneously used to erect the hotel block to first floor level and complete the steel transfer floor. The precast concrete crosswall frame would then be erected using a tower crane (luffing jib type) to avoid swinging over the adjacent main road.

At the tender stage, the precast concrete frame contractor had quoted on the basis of erecting the hotel using the contractor's tower crane. The precast contractor now offered an alternative price for using his own crane for the frame erection sequence – a 110 ton tracked low pivot jib mobile crane would be available for use. This option allowed a considerable cost saving when compared with the tender proposals. This option was proposed and accepted by all parties.

## Summary – value engineering

The value engineering proposals allowed a considerable reduction in the final bid price and secured the contract.

Value engineering can be defined as an organised approach to providing the necessary functions at the lowest possible cost. Value engineering plays an essential part when considering tender adjustments on a design and build contract.

Final tender adjustments at the contract negotiation stage often involve the principles of, "question everything in the original tender build-up." This often involves questioning design decisions made by the client's design team. In the case of the hotel and office block project both the foundation design and the steelwork were subject to redesign considerations at the value engineering stage.

Changes in methods of construction often come to the fore – for example, changes in the frame type or cladding type and their knock-on effects on other cost areas.

For a definition of value engineering see: www.mistronet.com

## Value engineering decision – method for supporting external facing brickwork

To support the external brick skin over the windows at each floor level, stainless steel angle brackets were bolted to the precast frame. This is considered an "expensive solution" in terms of both material and labour costs.

This proposal was considered a more economical option to providing a composite brick panel to form the external enclosure. The image show the brackets required to support the brickwork over the window openings to the external elevations. Stainless steel bracket supports are bolted to pre-formed brick lintels and bolted back to the precast frame.

**Brackets required to support the brickwork over the window openings**

## Completed facing brickwork to building elevations

**Brickwork in recessed panels above windows – an excellent quality finish**

The facing brickwork is held on steel angle supports fixed to the face of the precast concrete crosswall frame. The external facing task was completed on programme.

# Cranes – selection of crane for handling the PC frame erection

**Option one – tower crane proposal**

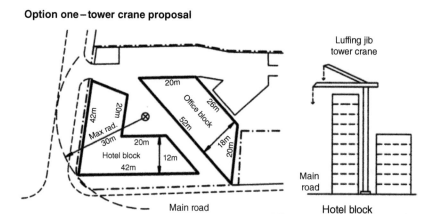

30m radius tower crane would
span onto roadway at front and
side of building

**Option two – mobile crane proposal – system adopted on site**

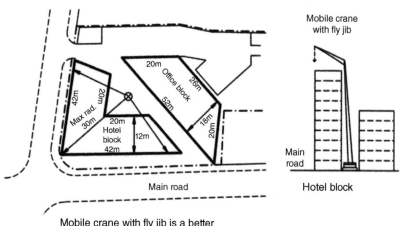

Mobile crane with fly jib is a better
choice and a more economical option

⊗ Crane location
Max. radius – 30 m

**Crane proposals**

## 8.9 Site layout planning

### General considerations

Every project has its own site layout problem, either too little space or too much (as in the case of this hotel and office project). Space available for site layout planning must be considered as a "flexible space". As a project progresses from the foundation stage to completion there is a change in the materials being stored and handled. The key to good site layout planning is therefore flexibility and thinking ahead.

Foundations and in-situ concrete frame construction create a need for space for fabricating reinforcement and formwork. Areas are also required for cleaning of the formwork and recovering usable materials. Access is required for ready-mixed concrete deliveries.

Steel-framed construction again requires extensive space for the storage of steelwork on the ground prior to lifting into position. Extensive space is also required for movement of plant and mobile lift platforms around the building.

When cladding operations commence, again space is required for glass storage frames and trestles. A clear space around the building is required for the movement of scissor lifts, hydraulic platforms and so on.

It is essential that consideration is made to placing materials on the floor space created within the building. This applies to items such as insulation materials, dry lining and plasterboard products, which need protection from the weather. Items such as bathroom pods may be positioned in each room during construction.

A forecast will be required of the number of office staff and operatives to be on site at peak times. This will influence the number of office units required and the canteen and drying rooms and toilet space to be provided for operatives and staff.

Site preliminaries encompass all areas of providing site facilities. This includes the provision of site accommodation, site services to maintain cabins, including electricity, erection and dismantling costs.

The estimator has to make a judgement at tender stage of the monies to be included in the estimate for such items. Average preliminaries allowances often equate to 6 – 12% of the contract sum.

It is necessary to keep a check during construction of the actual expenditure incurred in providing these services. Part of the monthly cost/value reconciliation process involves comparing the estimated allowances with the actual costs incurred. Many contractors overspend on contract preliminaries, especially when contract delays occur.

# Site layout plan – hotel and office project

**Extensive site compound and storage areas – central areas for vehicles moving materials to the building area**

**Contractor's main office set-up and subcontractors' storage areas**

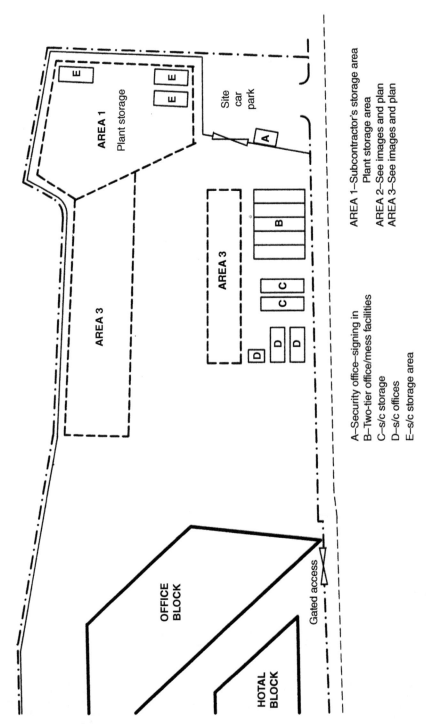

AREA 1
Plant storage

E

E E

Site
car
park

A

AREA 3

AREA 3

B

C C

D

D
D

A–Security office–signing in
B–Two-tier office/mess facilities
C–s/c storage
D–s/c offices
E–s/c storage area

AREA 1–Subcontractor's storage area
Plant storage area

AREA 2–See images and plan
AREA 3–See images and plan

OFFICE
BLOCK

HOTAL
BLOCK

Gated access

**Site layout proposals on proposed car park area at the rear of the development**

**AREA 2**

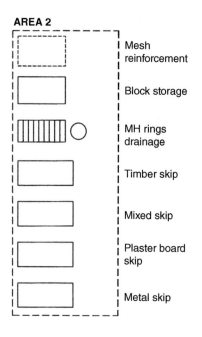

Mesh reinforcement

Block storage

MH rings drainage

Timber skip

Mixed skip

Plaster board skip

Metal skip

**AREA 3**

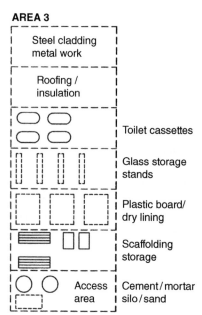

Steel cladding metal work

Roofing / insulation

Toilet cassettes

Glass storage stands

Plastic board/ dry lining

Scaffolding storage

Access area

Cement/mortar silo/sand

**Area 2 and area 3**

## Site offices and operatives facilities

**Two-storey office and mess**

**Project manager's office**

Office and mess facilities are provided in a two-storey block consisting of ten mobile units 9m×2m. The ground floor contains operatives' facilities such as canteen, drying rooms, toilets and store. The upper floor accommodates the site management personnel, including separate offices for the project manager, site manager, quantity surveyors, engineers and visiting personnel. A separate meeting room and samples area is also provided.

All site labour is directly engaged by the work package subcontractors. The main contractor specifies who is responsible for general movement of materials and maintaining a tidy site. Contra-charging of subcontractors applies when the subcontractor fails to leave a tidy workplace.

## Site offices and operatives facilities

**Excellent facilities provided in two similar canteen/mess areas, including microwave units and tea/coffee making**

**Food preparation area for the office staff – again similar to the facilities provided for the operatives**

**Area for displaying technical data and samples of products to be specified for the project**

## Materials management practices on site

A site foreman is in charge of the movement and storage of materials. This ensures an organised approach to materials management in order to minimise double handling. The images clearly indicate the importance of a tidy and clean site. Material management is a key feature of managing a successful project.

**Scaffold boards stacked under a wall scaffold area**

**Ducting stored at each floor level strapped to adjacent handrail**

**Glass storage racks for curtain walling material**

**Blocks stored on pallets at each floor level in the office block**

## 8.10 Hotel block – sequence of work for floors 5 to 7

High level floor

Low level floor

**Low level and high level floors**

Sequence of work per floor

Low-level floor: – 9 days per floor
High-level floor: – 7 days per floor

| Precast wall and floor sequence | | | | | | | | | | | | | | | | | | | |
|---|---|---|---|---|---|---|---|---|---|---|---|---|---|---|---|---|---|---|---|
| | | | | | | | | | May | | | | | | | | | | |
| Operation | Dur. | 2 | 3 | 4 | 5 | 6 | 7 | 8 | 9 | 10 | 11 | 12 | 13 | 14 | 15 | 16 | 17 | 18 | 19 |
| 4th floor | | | | | | | | | | 4th floor complete | | | | | | | | | |
| Erect walls | 5 days | | | | | | | | | | | | | | | | | | | |
| Erect floor | 2 days | | | | | | | | | | | | | | | | | | | |
| | | | | | | | | | | | | | | | | | | | | |
| 5th floor | | | | | | | | | | | | | | | | | 5th floor complete | | | |
| Erect walls | 5 days | | | | | | | | | | | | | | | | | | | |
| Erect floor | 2 days | | | | | | | | | | | | | | | | | | | |
| | | | | | | | | | | | | | | | | | | | | |

**Programme for high-level floor**

The work package subcontractor is responsible for liaising with the precast supplier. An agreed sequence of work per floor requires units to be delivered to suit the erection sequence.

## Typical floor plan – hotel block

This shows 200 mm thick crosswall layout.

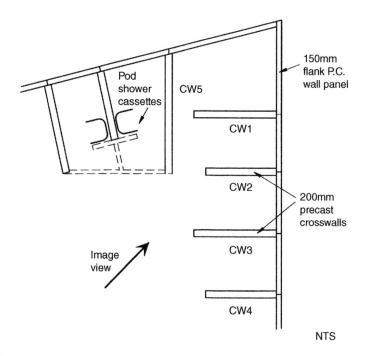

**Flank walls erected and propped – internal division walls stabilise the frame**

**Crosswalls at 2.8 m centres forming a single bedroom space**

## Hotel block images

**Progress at fifth floor of crosswall frame – note the handrail fitted to the top edge of the external wall units**

**200 mm thick precast crosswall held in vertical position with extendable metal props**

**Bathroom shower cassettes lifted into rooms as work progresses – temporary propping support shown to external wall panels**

## 8.11 Programme for hotel building

### Precast crosswall frame and steelwork to first floor

An elevation of the hotel block is shown, indicating the ground floor and mezzanine retail area with the six-storey hotel above the transfer floor level. The low-level area contains three floors of bedroom units with three further floors forming the high-level unit.

A draft pre-contract programme is shown which includes the steel frame and transfer floor for the retail unit, together with the crosswall construction for the hotel.

The cycle times for the erection of the precast crosswall frame per floor:

Low-level frame: – 9 days per floor
High-level frame: – 7 days per floor

Based on the bar chart programme sequence, an overall construction period of 78 days has been assessed for the complete hotel block. This includes a 30 day period to construct the steelwork and the transfer floor to the retail area.

An extract from the project programme prepared by the project manager is shown. This includes the mobilisation and removal of the tracked crawler crane on completion of the work, an overall programme period of 94 days.

The construction of the crosswall precast frame was a complex operation due to four suppliers being involved in providing the precast components. This involved preparing a daily schedule of precast unit deliveries from four suppliers. Daily precast delivery schedules were prepared for each floor and just-in-time scheduling applied.

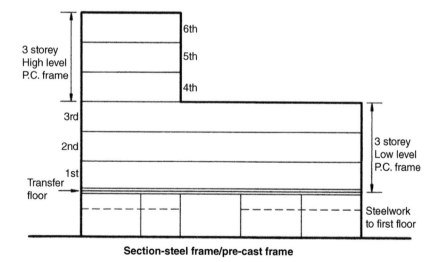

**Section-steel frame/pre-cast frame**

**Section – steel frame/ pre-cast frame**

During the erection of the frame, good contractor, supplier and subcontractor relationships ensured that the frame erection programme was successfully achieved.

## Planned dates for crosswall frame erection

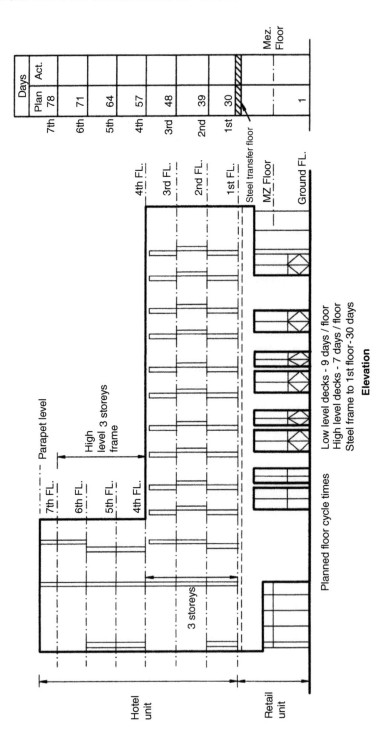

**Elevation**

**Dates for erection of crosswall frame**

## Programme for transfer floor and precast crosswall frame

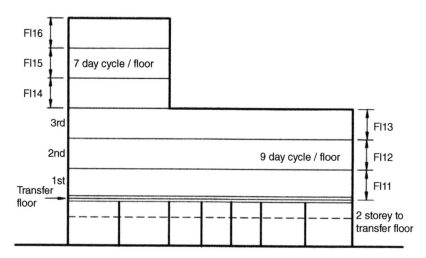

**Transfer floor and precast crosswall frame programme**

## Programme assessment

| | |
|---|---|
| Mobilisation at commencement of work | – 10 days |
| Preparation of area and erection and location of crane | |
| Steelwork to transfer floor level at precast deck | – 30 days |
| Lower storeys – PC crosswall frame three floors at build rate of 9 days per floor | – 27 days |
| Upper storeys – PC crosswall frame three floors at build rate of 7 days per floor | – 21 days |
| Dismantle crane on completion | – 6 days |
| Clear site area | |

**Total 94 days**

## Planning sheet

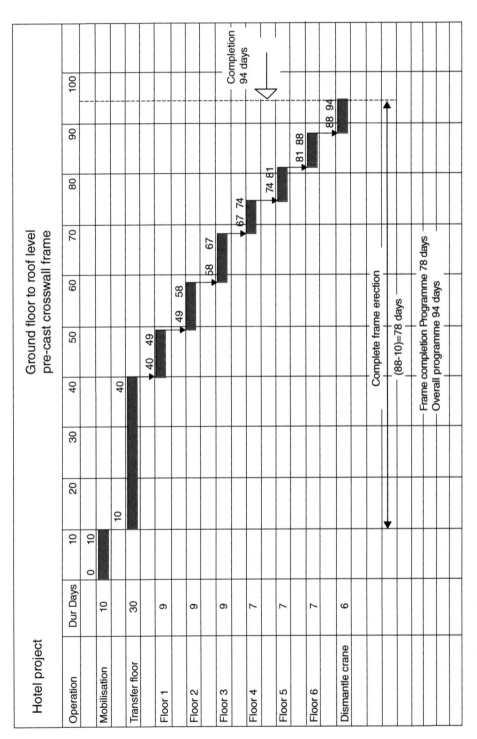

**Hotel project planning sheet**

## 8.12    Site monitoring of external brickwork operations

The contractor proposed the use of a separate external skin of blue engineering brick, in lieu of the composite wall panel indicated on the building elevations. The project manager proposed to use a colour-coded progress recording system to indicate the progress per week at each stage of work.

The scheduled forecast dates for the completion of brickwork per lift is illustrated on the hotel elevation drawing. A forecast of the brickwork completion dates for the first, second, third and fourth floors is indicated in the planned date schedule. Actual completion dates will be monitored alongside each key stage.

A colour coded record diagram of the actual brickwork progress is shown on the road elevation. Actual progress is shown in red, up to 11/4 and yellow to 18/4.

The colour coding of drawings (building elevations and plans) is an excellent way of monitoring progress. This clearly shows, pictorially, the relationship between planned and actual progress. Similar colour coding may be used for recording progress to floor beams, column sequences and the pouring of the floor bays. This is an ideal task for an assistant manager or trainee site engineer.

**Elevation of completed building**

## Planned dates for brickwork progress

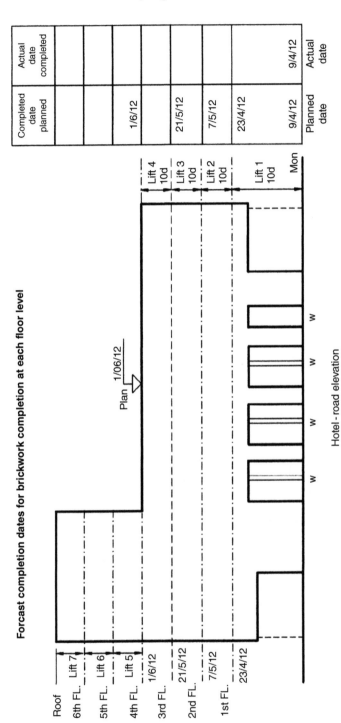

Forcast completion dates for brickwork completion at each floor level

| | Completed date planned | Actual date completed |
|---|---|---|
| | | |
| | | |
| | 1/6/12 | |
| | | |
| | 21/5/12 | |
| | 7/5/12 | |
| | 23/4/12 | |
| | 9/4/12 | 9/4/12 |
| | Planned date | Actual date |

Plan 1/06/12

Lift 4 10d
Lift 3 10d
Lift 2 10d
Lift 1 10d

Mon

Roof
6th FL. Lift 7
5th FL. Lift 6
4th FL. Lift 5
1/6/12
3rd FL.
21/5/12
2nd FL.
7/5/12
1st FL.
23/4/12

Hotel - road elevation

**Planned dates**

## Actual dates for brickwork progress

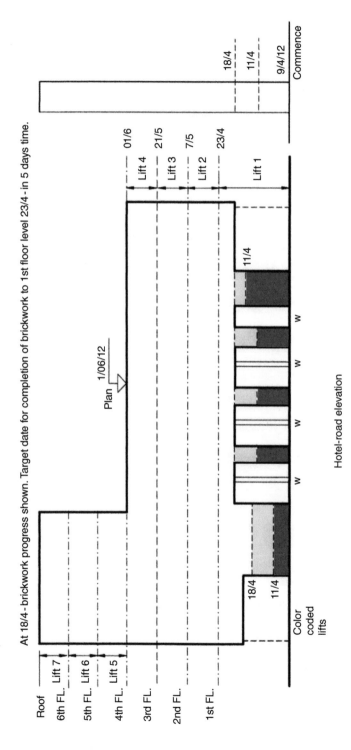

## 8.13   Office block – programme of work exercise

**Steel frame with metal deck floors**

**Rear view with external cladding in progress – plant room completed at roof level**

**Glass curtain walling in progress to side elevations**

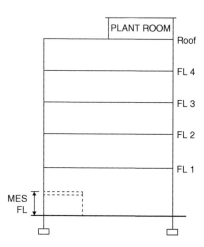

**Six storey building**

# Office block – programme assessment sheet

| Element | Description of work | Duration |
|---|---|---|
| 1. Foundations | Bullivant NRG dynamic compaction system used to stabilise area. Pad foundations and ground beams constructed to form foundations to frame. | 1 week NRG<br><br>3 weeks Fdts |
| 2. Erect steel frame | Five-storey steel frame erected plus plant room on part of roof area. | 6w. steel frame |
| 3. Internal floors<br><br>Metal deck<br><br>Concrete | Holorib metal deck floors fixed to steelwork.<br><br>Access formed via precast concrete stairs installed with frame erection.<br><br>150 mm in-situ concrete slab cast on metal deck. | 4w. fix metal deck<br><br>4w. concrete 5 floors |
| 4. Plant room | Erect steelwork to plant room. | 1 week |
| 5. External roof (building watertight) | 150 mm insulation fixed on top of in-situ slab, laid to falls. Trocal single layer finish. | 1 week |
| 6. Curtain walling | Aluminium curtain wall framework fitted to three elevations of building. Rear wall in rainscreen cladding. | 14 weeks |
| 7. Building services and internal partioning | Installation of building services/ internal partitioning on each floor level with block walls around staircase and lift areas. | 29 weeks |
| 8. External works | Paving around buildings and construction of 40-bay car parking facility. | 16 weeks |

**Programme assessment sheet**

## 8.14 Student programming task

Using linked bar chart principles, develop a pre-construction bar chart programme, based on the data shown on the programme assessment sheet.

**Notes:**

Think about the linked bar chart principles. Refer to Chapter 9, p145, of *Construction Planning, Programming and Control*, which illustrates the relationships used when preparing a linked bar chart. This applies to the preparation of a hand-drawn bar chart or when using project planning software such as Asta Development's Powerproject or Teamplan.

Think logically:

- Which operations must finish before the next one commences (finish-to-start relationship)?
- Which operations can start together (start-to-start relationships)?
- Which operations can overlap each other (start-to-start relationships)?
- Which operations finish at the same time or have a relationship between finish dates.

**Think and plan:**

Plan out the bar chart in a draft on a simple blank programme sheet, with durations shown in weeks. Access the overall project period for the office project (approximately 59 weeks, depending on your logic).

Reference – *Construction Planning, Programming and Control*, Cooke & Williams, 3rd Edition, Blackwell Publications

## Planning sheet

| Operation | Dur weeks | 10 | 20 | 30 | 40 | 50 | 60 | 70 | Start | Finish |
|---|---|---|---|---|---|---|---|---|---|---|
| Foundation | 4 | 0–4 | | | | | | | 0 | 4 |
| Erect steel floor | 6 | 4–10 | | | | | | | 4 | 10 |
| Internal floors-metal deck | 4 | 10–14 | | | | | | | 10 | 14 |
| Internal floors-concrete | 4 | 11–15 | | | | | | | 11 | 15 |
| Plant room | 1 | 16 | | | | | | Week 59 | | 15 | 16 |
| roof | 1 | 17 | | | | | | | | 16 | 17 |
| Curtain walling | 14 | 15–29 | | | | | | | 15 | 29 |
| Building services | 29 | | | 27–56 | | | | | 27 | 56 |
| Internal work | | | | | | 43–59 | | | | |
| External works | 16 | | | | 43–59 | | | | 43 | 59 |

Office block

First attempt

**Office block planning sheet**

## 8.15 Completed office building

**Six storey steel frame office building with front and side elevation finished in feature yellow panels, and two-storey retail unit to ground floor area**

**Completed office building**

## IBIS hotel and office project

**View from the road elevation of the hotel and office block complex**

**Five to eight storey hotel block finished in blue engineering bricks**

# Chapter 9

# The Merlin Project

## Contents

| | | |
|---|---|---:|
| 9.1 | Contents of the case study | 144 |
| 9.2 | Project overview | 144 |
| 9.3 | Project information | 146 |
| 9.4 | Client/contractor relationships | 146 |
| 9.5 | Head office and company organisation structure | 147 |
| 9.6 | Management structure and site organisation | 148 |
| 9.7 | Site layout plan - site compound area | 149 |
| 9.8 | Programming the project | 152 |
| 9.9 | Sequence of construction of the main portal frame | 156 |
| 9.10 | Erection sequence for the main steel frame | 157 |
| 9.11 | External building features | 161 |

*Management of Construction Projects*, First Edition. Brian Cooke.
© 2015 John Wiley & Sons, Ltd. Published 2015 by John Wiley & Sons, Ltd.

## 9.1 Contents of the case study

**Elevation of the completed building**

This case study covers the construction and management aspects of the building of a project in rural Derbyshire. The ten month project was undertaken by a medium sized local contractor involving thirty five work package subcontractors. The contract was successfully completed within the planned contract period.

The following aspects of the project are outlined:

- Client/contractor relationships.
- An overview of the contractors head office organisation and the site management structure.
- The contractors approach to the development of a work sequence for the main steel frame.
- An overview of the steelwork and services procurement programme.
- A detailed analysis of the sequences involved in erecting the 40 m span portal frames to the main factory building.
- The construction approach taken to the provision of limestone filled gabion wall panels to various elevations.

## 9.2 Project overview

The famous "Buxton Water" brand has been established since the early 1800's. The project involves the building of a new factory/ head office and bottling plant capable of producing one billion bottles per year.

The building is required to blend into the Derbyshire countryside, with external finishes including limestone filled gabions and gritstone walls.

Extensive external works provide an on-site reservoir which allows water to be filtered into the limestone rocks below and a foul water treatment plant is to be provided. Natural spring water is to be piped to the plant from two local sources located some 4 kilometres from the project.

The complex includes the main factory building (bottling plant), a water treatment building and a two storey office block. The site plan illustrates the layout of the project.

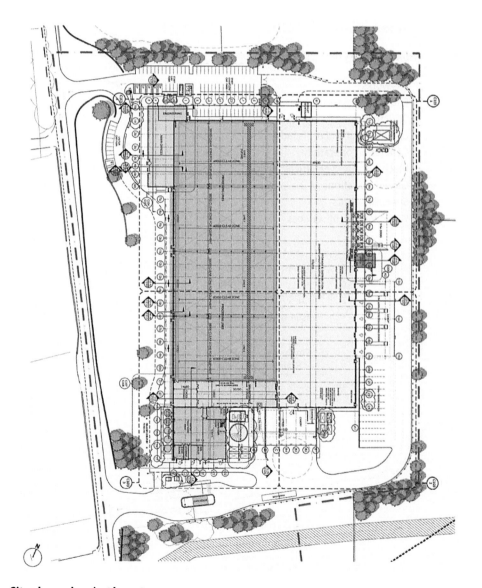

**Site plan and project layout**

## 9.3 Project information

| | |
|---|---|
| Client: | *Buxton Water (part of the Nestle group)* |
| Contractor: | *Pochin Construction* |
| Contract value: | *£15.5 million* |
| Contract period: | *9 months (40 weeks)* |
| Commencement date: | *August 2011* |
| Form of contract: | *Design and Build (with novation)* |
| Architect: | *Fletcher-Rae (scheme design)* |
| Employers Agent: | *Spring and Company* |
| Consulting Engineer: | *Scott Hughes* |
| Services Engineer: | *QED Ltd* |

Work Package Contractors (35 in total):

- Siteworks
- Foundation bases
- Structural steel frame
- External wall and roof cladding
- Concreting operations
- Mechanical and electrical services
- Glazed elevations
- Specialist doors
- Water treatment services

## 9.4 Client/contractor relationships

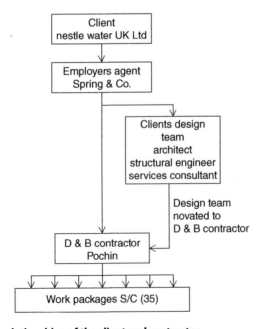

**The structure of the relationships of the client and contractor**

## 9.5 Head office and company organisation structure

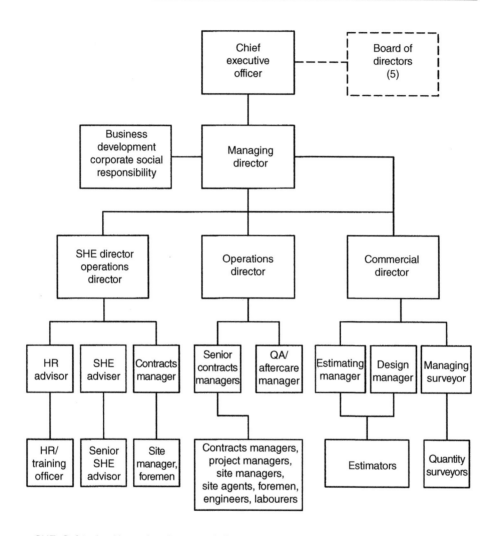

SHE- Safety, healthy and environmental director

HR- Human resources advisor

**The structure of the head office and company organisation**

Pochin contractors have been a well established north-west contractor since the 1950's and family members are still involved in the management of the company and its subsidiaries.

It is interesting to note that the titles of Contracts Manager, Project Manager, Site Manager, Finishing Foreman and Foreman are still included in the company organisation structure. This does not apply to many of today's larger contractors.

## 9.6 Management structure and site organisation

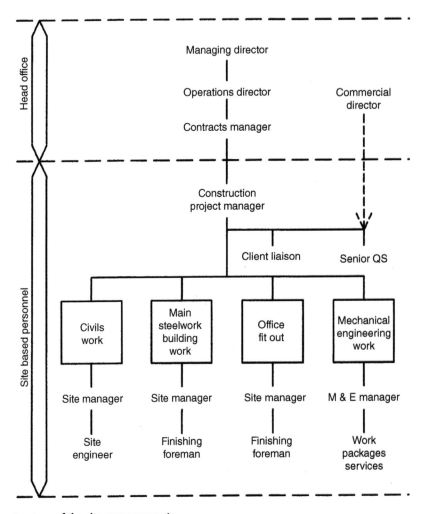

**The structure of the site management**

## 9.7 Site layout plan - site compound area

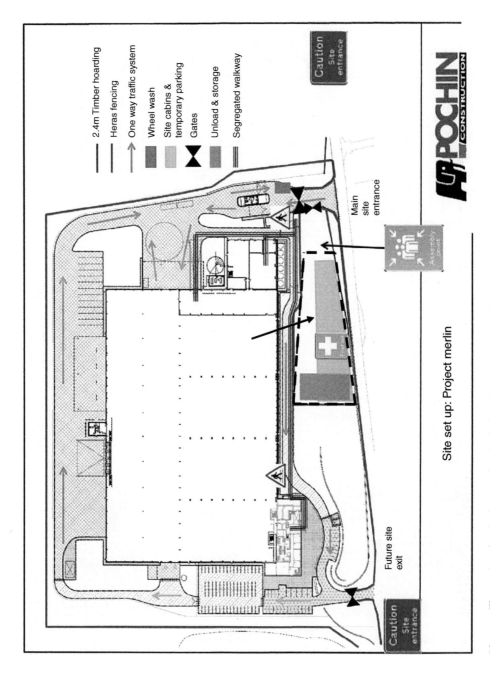

**Diagram illustrating the site layout plan, including the site compound area**

## Site compound layout

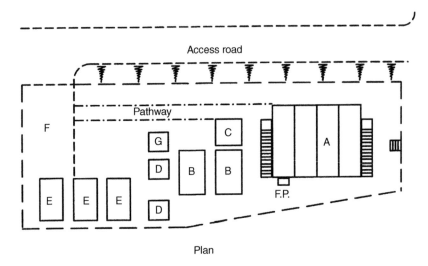

Plan

**Plan of the site compound**

Key:

A: 2 Tier offices
B: Subcontractors
C: Outdoor smoking
D: Storage cabins
E: 2 & 3 Tier S/C offices
F: Material storage areas
G: Hot food servery

**Site material storage**

**Hot food servery**

## Site accommodation: block A

Eight tiered fully serviced mobile offices making up the main contractor's and design team accommodation

Site covered smoking area plus site canteen unit serving hot food

## Site accommodation: subcontractors block B

Main Work Package subcontract accommodation (location E)

**Commencement of concrete pumping to rear access road area**

## 9.8 Programming the project

A linked bar chart illustrates the main phase of the work for the 40 week project

Reference must be made to the site plan and project layout on page 145. The sequence of work is as viewed from the main road fronting the project and working from right to left.

Work sequences for erecting the steel frame:-

Steelwork erected to the water treatment plant(shaded blue)

- Construction of the highbay portal (shaded pink)
- Erection of the four bay 40m span portals (shaded blue)
- Steelwork to the office block (shaded green)
- Steelwork to the near bays (shaded pink)

The linked barchart for the main phases of work is illustrated

Bar chart displays are illustrated for the steelwork procurement and also for the site electrical services.

The project planning was under the direct control of the project manager aided by an experienced site based planning engineer.

Once again, the Asta Development Power Project package was utilised on the project.

The sequence of work from the high bay portal and the flour 40m span bays to the main factory unit are illustrated both pictorially in bar chart format.

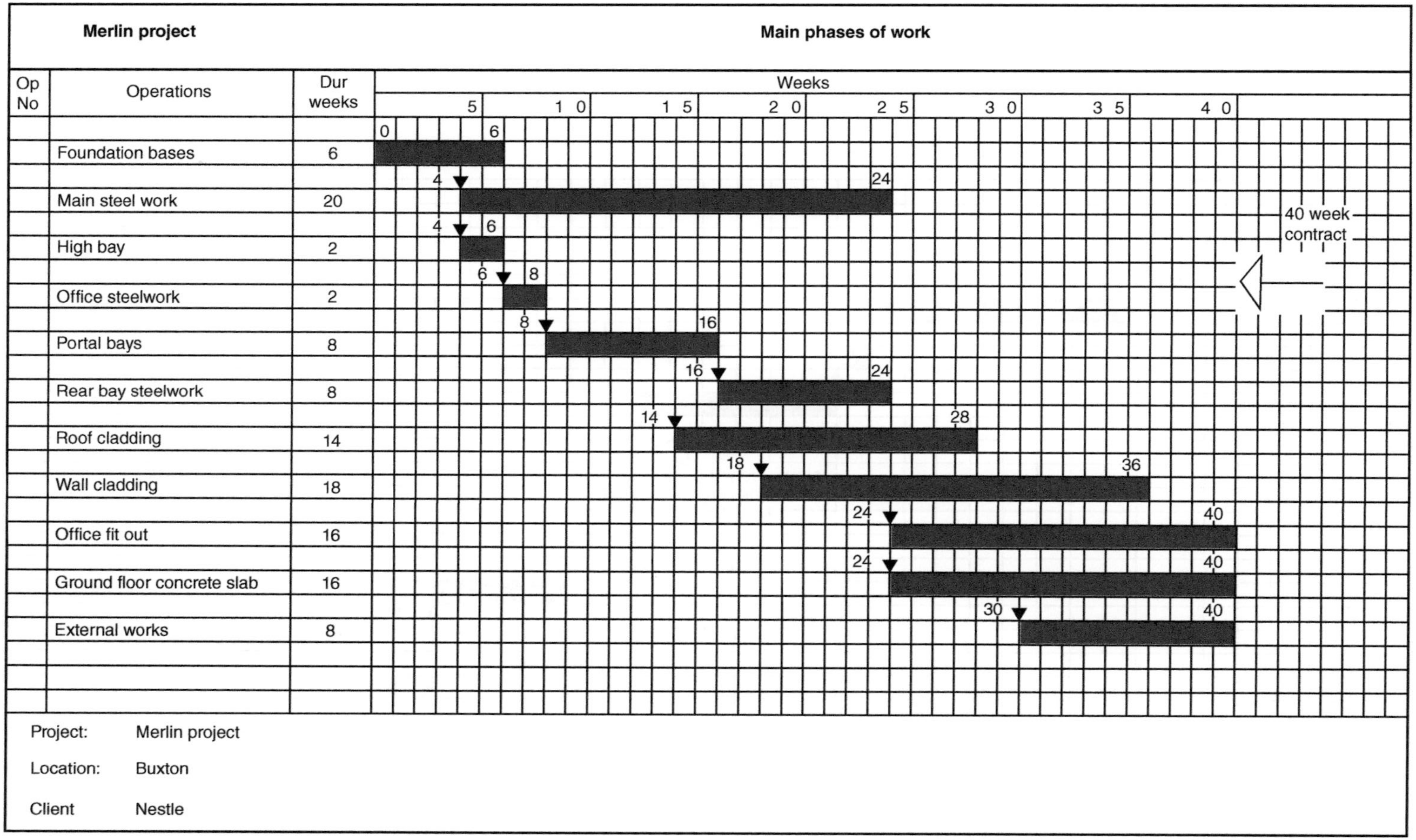

# Main phases of work

**Merlin project**

| Op No | Operations | Dur weeks |
|-------|------------|-----------|
| | Foundation bases | 6 |
| | Main steel work | 20 |
| | High bay | 2 |
| | Office steelwork | 2 |
| | Portal bays | 8 |
| | Rear bay steelwork | 8 |
| | Roof cladding | 14 |
| | Wall cladding | 18 |
| | Office fit out | 16 |
| | Ground floor concrete slab | 16 |
| | External works | 8 |

Weeks: 0, 5, 10, 15, 20, 25, 30, 35, 40

40 week contract

Project: Merlin project
Location: Buxton
Client: Nestle

**Main phases of work**

A procurement programme is illustrated for the steelwork package together with a services procurement programme.

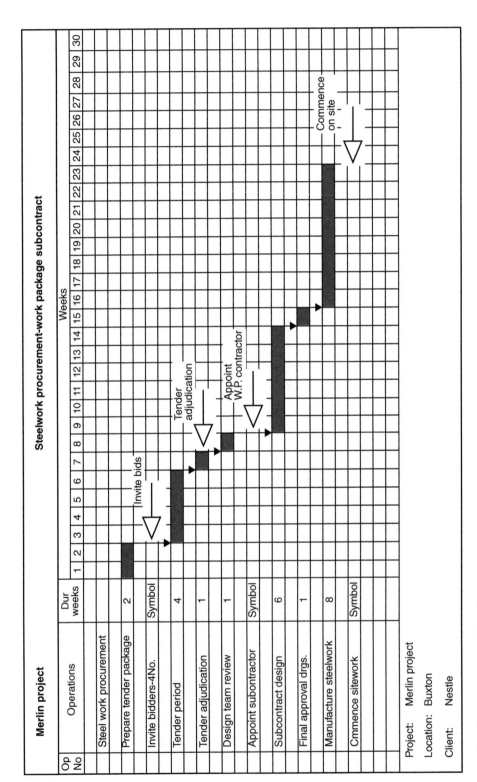

**Steelwork procurement work package**

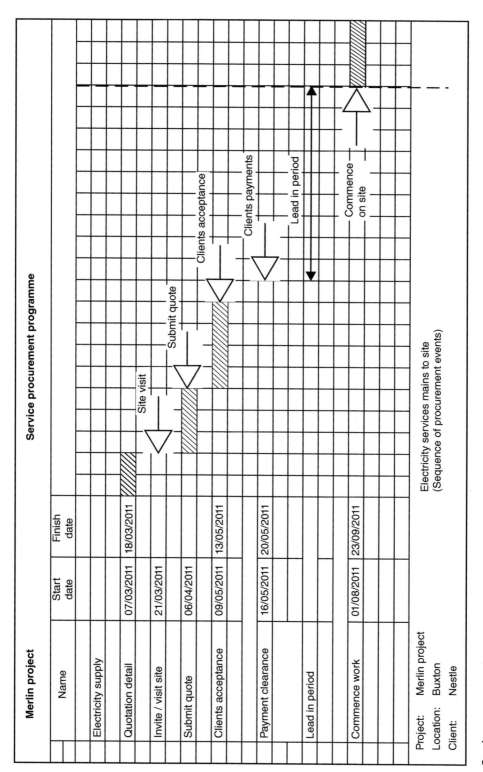

**Merlin project**                                          Service procurement programme

| Name | Start date | Finish date | | |
|---|---|---|---|---|
| Electricity supply | | | | |
| Quotation detail | 07/03/2011 | 18/03/2011 | | Site visit |
| Invite / visit site | 21/03/2011 | | | Submit quote |
| Submit quote | 06/04/2011 | | | Clients acceptance |
| Clients acceptance | 09/05/2011 | 13/05/2011 | | Clients payments |
| Payment clearance | 16/05/2011 | 20/05/2011 | | Lead in period |
| Lead in period | | | | Commence on site |
| Commence work | 01/08/2011 | 23/09/2011 | | |

Project:  Merlin project          Electricity services mains to site
Location: Buxton                  (Sequence of procurement events)
Client:   Nestle

**Service procurement programme**

## 9.9 Sequence of construction of the main portal frame

The main factory unit consists of an 18 metre span portal plus four 40 metre span portal bays. The construction sequence is illustrated for each of the five portals together with a series of photographic images recorded on site.

On completion of the five portals, the steel erection continued to the two bay distribution warehouse area to the rear elevation. Work continued on the main factory unit in order to form the wave roof structure. On completion of the fixing of the roof purlins, work could commence on the roof insulation decking.

The wave roof forms a key architectural feature to the front elevation of the building.

### Steelwork erection

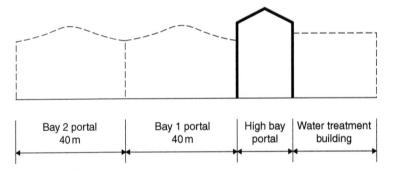

| | | | |
|---|---|---|---|
| Bay 2 portal 40 m | Bay 1 portal 40 m | High bay portal | Water treatment building |

**High bay portal figure**

| High bay programme | | | | | | | | | | | | | | | | | | | | | | |
|---|---|---|---|---|---|---|---|---|---|---|---|---|---|---|---|---|---|---|---|---|---|---|
| Programme-as built record | | | | | | | | | | | | | | | | | | | | | | |
| Operations | Duration | 1 | 2 | 3 | 4 | 5 | 6 | 7 | 8 | 9 | 10 | 11 | 12 | 13 | 14 | 15 | 16 | 17 | 18 | 19 | 20 | |
| Eract frame | 5 days | ///////////// | | | | | | | Start day 1 complete day 5 | | | | | | | | | | | | | | |
| Erect roof Purlins | 4 days | | | | ///////////// | | | | Start day 4 complete day 7 | | | | | | | | | | | | | | |

**High bay programme**

The sequence of work for the main factory unit commences with the construction of a high bay portal adjacent to the gable of the water treatment building. This is followed by the erection of the four bay 40 m span portals forming the remainder of the main factory unit. On completion of the main factory unit, the construction of the two bay portal framed rear main building commenced.

## 9.10   Erection sequence for the main steel frame

### High bay portal

**Commencement of the erection of the nine bay high level portal frame**

**Completed high bay portal**

## Steelwork erection

### Portal bay 1

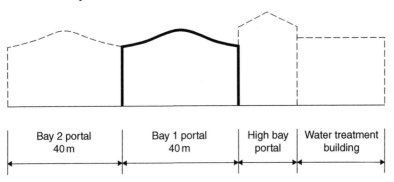

| Bay 2 portal 40 m | Bay 1 portal 40 m | High bay portal | Water treatment building |

**Bay 1 portal**

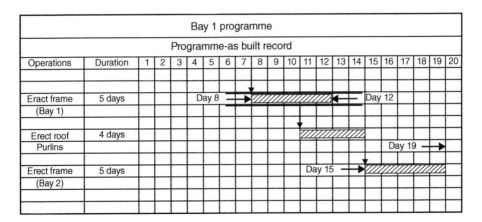

**Bay 1 programme diagram**

Portal Frame Bay 1: commence Day 8, complete Day 12
Portal Frame Bay 2: commence Day 15, complete Day 19
Portal Frame Bay 3:commence Day 22, complete Day 26

Each of the four main frames commences at 7 day intervals

**Commencement of bay 1**

## Portal bay 2

**Commencement of bay 2: gable frame**

**Erection of bay 2 in progress**

## Portal bays 2, 3 and 4

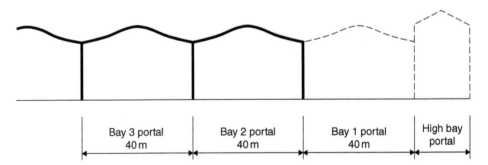

| | | | |
|---|---|---|---|
| Bay 3 portal 40 m | Bay 2 portal 40 m | Bay 1 portal 40 m | High bay portal |

**Portal bay 2, 3 and 4**

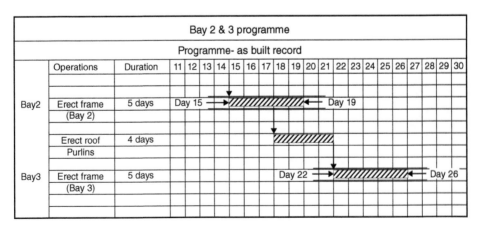

| | Bay 2 & 3 programme | | | | | | | | | | | | | | | | | | | | | | |
|---|---|---|---|---|---|---|---|---|---|---|---|---|---|---|---|---|---|---|---|---|---|---|---|
| | Programme- as built record | | | | | | | | | | | | | | | | | | | | | | |
| | Operations | Duration | 11 | 12 | 13 | 14 | 15 | 16 | 17 | 18 | 19 | 20 | 21 | 22 | 23 | 24 | 25 | 26 | 27 | 28 | 29 | 30 |
| Bay2 | Erect frame (Bay 2) | 5 days | Day 15 → ///////// ← Day 19 | | | | | | | | | | | | | | | | | | | | |
| | Erect roof Purlins | 4 days | ///////// | | | | | | | | | | | | | | | | | | | | |
| Bay3 | Erect frame (Bay 3) | 5 days | Day 22 → ///////// ← Day 26 | | | | | | | | | | | | | | | | | | | | |

**Portal bay 2 and 3 programme**

**Wave roof to bay 1**

**View of wave roof as purlins and overhang are completed**

## 9.11 External building features

In order for the building to blend into the Derbyshire landscape, limestone filled gabions have been introduced into the building elevations. Gabion panels have been introduced between the main portals on the front elevation of the building. The gabions have been filled manually with limestone from a scissor lift. Additional low gabion walls have also been constructed around the factory building.

The whole elevation of the office block is faced with 300 mm wide gabion cages which are held in position on vertical aluminium support rails. The filling of the gabions is carried out by hand, from an hydraulic platform, which appears to be an expensive labour solution to the filling of the gabions.

## Main building and office block

**Cladding to office block elevation in progress, stone gabions to form wall feature to front face of office block and returns**

**Feature gabion panels between portal bays, gabions filled with stone from scissor left platform**

## Gabion feature walls to office block and factory

**Front elevation of main factory**

## Sequence of work in constructing gabion faced feature walls to office block

The building design is intended to blend in with the natural Derbyshire surroundings of the Peak district.

The gabion walls are held in position on vertical aluminium support rails. The gabions are filled by hand from an hydraulic scissors lift platform.

The tasks involved are very manually oriented and expensive in labour cost.

**Filling containers with limestone from scissor platform**

**Filling gabion cages by hand in final position**

**Transport of scissors lift platform to wall location**

**Vertical support rails for tying gabions back to wall**

# Chapter 10

# The Co-operative Head Office Building

## Contents

| | | |
|---|---|---|
| 10.1 | Contents of the case study | 166 |
| 10.2 | Project overview | 166 |
| 10.3 | Project information | 167 |
| 10.4 | Client/contractor relationship | 168 |
| 10.5 | Contractors senior project management structure | 169 |
| 10.6 | Site management structure | 170 |
| 10.7 | Construction stages of the building frame | 171 |
| 10.8 | Managing the surveying function | 174 |
| 10.9 | Managing the design process | 176 |
| 10.10 | Managing the site logistics | 178 |
| 10.11 | Sequence of erecting the steel frame and precast floors | 183 |
| 10.12 | Managing the building enclosure | 189 |

*Management of Construction Projects*, First Edition. Brian Cooke.
© 2015 John Wiley & Sons, Ltd. Published 2015 by John Wiley & Sons, Ltd.

## 10.1 Contents of the case study

**The completed building**

This is one of the most prestigious building projects undertaken in Manchester over the past 10 years. The building is an outstanding design and creates a stunning structure within the business centre of the city.

This case study encompasses the following aspects of the project:

- The project overview outlining the buildings technological features.
- An overview of the client/contractor relationship and the contractors approach to managing the contract.
- The main construction stages of work are outlined for the main four stages of work.
- The management aspects of the project relating to the management of the surveying function, the design process and the site logistics.
- The sequence of the erection of the main steel frame and the management of the building enclosure.

## 10.2 Project overview

The new head office building will create one of the most sustainable commercial buildings in Europe. It is one of the first in the UK to be built to meet the BREEAM outstanding designation standards with a score of 95.3.

The building will have its own source of sustainable heat and power generation using bio fuel and waste cooling oil. The building has been designed as a zero carbon emissions building and designated as an energy plus building.

Four and a half acres of the site have been developed as a public space.

The key dates in relation to the project are as follows:

Tender date: *Early 2010 (appointed in June)*
Work commenced: *July 2010*
Work completed: *March 2013*
Building occupied: *November 2013*
Cost: *£195 million*
Form of contract: Design and Build

Eco- friendly features include:

- Heat recovery system from the atrium and IT systems to heat the building.
- Combined used water and waste recycling.
- Low water consumption appliances.
- High efficiency lifts.

Other building facilities include:

- A café within the atrium provides a communal focus.
- A restaurant area on the eighth floor provides extensive city views.
- 320 person auditorium at ground floor level.
- Full disabled access.
- Extensive parking for 150 cars and 105 bicycles.

The cost budget for the final project was forecast within the range of £2200 and £2500 per square metre.

## 10.3   Project information

### Project particulars

Client:                              *The Co-operative Society*
Contractor:                          *BAM Construction Group*
Contract value:        *£117 million*
Contract period:    *37 months*
Start date:              *Enabling works –July 2010*
Form of contract:   *Design and Build*

Architect:                                   *3D Reid*
Consulting Engineer:                         *Buro Happold*
Project Managers:                            *Gardiner & Theobald*
Building Services Consultant:   *SI Seeley*
Building Control:                            *Manchester Council*

Major work packages:

| | |
|---|---|
| Basement and staircase towers: | *P C Harrington* |
| Steelwork: | *Fisher Engineering* |
| External cladding: | *Wargner Buro* |
| Coffered floors: | *McCory Construction* |
| Building Services: | *Rotary North West/ Sol Acoustics* |
| Roofing: | *Helix Roofing* |

## 10.4   Client/contractor relationship

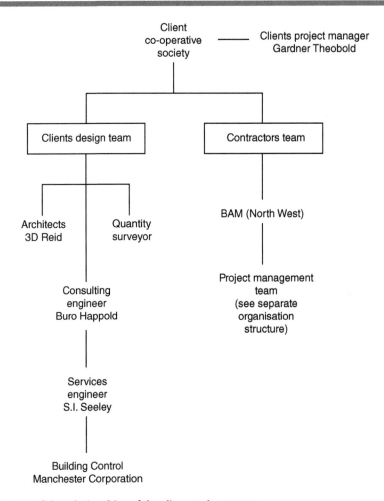

**The structure of the relationships of the client and contractor**

## 10.5 Contractors senior project management structure

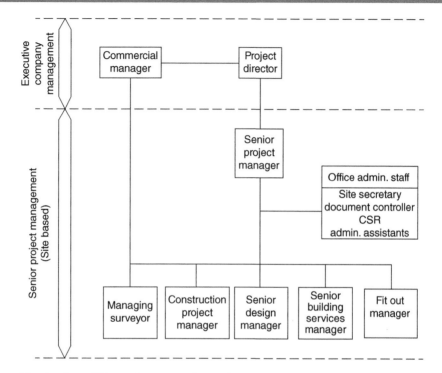

**The structure of the contractors senior project management team**

**Four storey office / mass building**

## 10.6 Site management structure

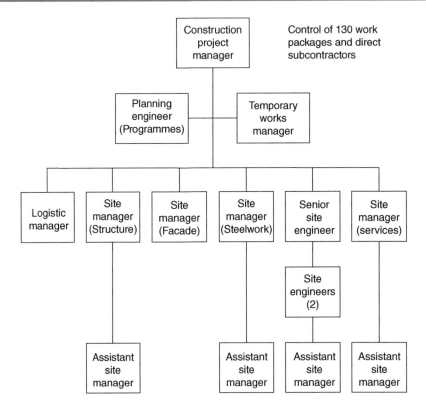

**Organisation of project work packages**

**BAM Project information board**

## 10.7   Construction stages of the building frame

The construction stages of work on the building frame consist of:

- Basement construction.
- Staircase cores.
- Steelwork and floors.
- Building enclosure – double skin façade.

A brief description of each of the above stages is outlined, supplemented by site images.

### The basement construction

The basement is surrounded with a concrete bored pile wall containing some 2000 piles. The two storey deep basement contains insitu concrete columns supporting flat slab insitu floors up to the ground floor level.

A centrally located schwing concrete mast pump was used for all the concreting operations. The logistics in relation to the pump location and concrete deliveries is outlined in the management section.

**View of SCHWING mast pump (P C Harrington – work package contractor**

**Mast pump located at ground floor level (placing concrete in the basement area)**

## The staircase cores

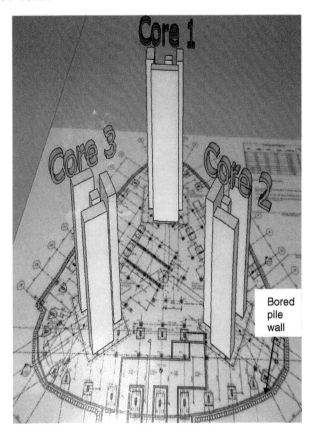

**Staircase cores**

The work package for the basement and staircase cores was undertaken by P C Harrington, a specialist package contractor. The three cores in each of the towers was constructed using the slipform formwork system. Concrete pumping was maintained to the 12$^{th}$ floor level, the final three floors being placed by tower crane and skips.

**Image of staircase cores**

Cores 1, 2 and 3 consist of three staircase towers linked by flank walls and beams. The average construction period per core is approximately 6 to 7 weeks. The concrete cores assist in providing stability ties to the steel frame at each floor level.

## Multi-story steel frame

A sequence study is later illustrated for the erection of the sixteen storey frame.

Floors 1 and 2 above the basement level consisted of a series of precast perimeter columns supporting the second floor steelwork.

Steelwork to floors 3, 4 and 5 was then erected incorporating the precast coffered floor units. Subsequently floors 6, 7 and 8 followed. Above floor 8, the building frame is stepped back to incorporate the vierendeel steel frames. These form a feature on the front elevation of the building.

**Elevation of floors 9 to 15 indicating vierendeel trusses**

## Building the double skin façade

The sequence of fixing the vertical glazed wall panels to the face of the building frame are illustrated later. The method chosen at tender stage eliminated the need for an external scaffold as the glazed panels could readily be fixed from inside the building.

Secondary over cladding was supported on an independent framework to provide a double skin façade - this forms one of the main design features of the completed building.

## 10.8  Managing the surveying function

The responsibility of the surveying function within each of the companies regions is controlled by the Commercial Surveying Director based in the company's regional offices. He is responsible for appointing the site based surveying team under the control of the managing surveyor based on site.

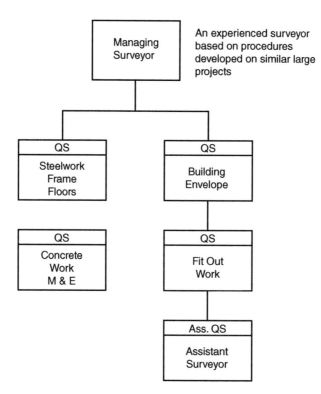

**Organisation of the surveying function**

The managing surveyor works closely with the procurement team led by the senior design manager/ procurement manager.

Responsibilities of the managing surveyor include:

- Liaison with the senior project manager responsible for the project.
- Liaison with the procurement manager's with regards to the placing of work packages and domestic subcontracts.
- Organising the surveying team under his control.
- Direct responsibility for the monthly cost/ value report – this includes forecasting cost and value to project completion.
- Liaison with the clients quantity surveyors regarding the monthly valuations and variations to the contract.
- Establishing data on contract claims and extensions of time.
- Reporting to both the senior project manager and commercial surveying director on the contract surveying situation.
- Control and payment of monthly valuations for all work packages and domestic subcontractors.
- Settlement of final accounts with all subcontractors as the contract progresses.

Good relationships between the main contractor, the work package subcontractors and domestic subcontractors is essential throughout the contract. Disputes between parties should not be allowed to disrupt the progress of work on site.

## 10.9   Managing the design process

It is essential to control the flow of information from the clients design team and the work package contractor providing a design package service. The responsibility on the project lies with the senior design manager.

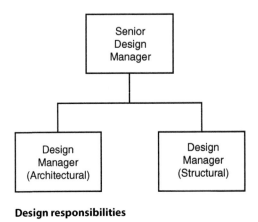

**Design responsibilities**

A schedule of design responsibilities has been produced to clarify the responsibilities for design within each work package particularly with respect to contractor design portions. These have been separated into three design areas:

Type A:  *Contractor Design*
Type B:  *Consultant Design with some Contractor Design*
Type C:  *Full Consultant Design*

### Documentation for the co-ordination of the contractor design portion of the contract

A schedule has been produced by the Senior Design Manager to clarify the responsibility for design within each work packages. This relates specifically to the contractor design portion.

The responsibility for design across the work packages (subcontractor packages) has been divided into three types:

### Type A: contractor design

This design will be completed by the contractor based upon the Employers Requirements issued by the consultants Employment Requirements and will define the scope of the work, key dimensional criteria, overall required appearance including materials and colour. The Employers Requirements also specify the criteria to which the design must perform, set standards and the quality of materials, components, assemblies, fabrication, manufacture and testing requirements. The Employers Requirements will also define how the works are co-ordinated with other packages, irrespective of the design responsibility to the package works. Type A work packages which fall into this category include:

Type A work packages which fall into this category include:

- Precast concrete coffered floor slabs.
- Precast stairs and landings.
- Curtain walling.
- Precast wall panels and stone cladding.
- Main safes.
- Window cleaning cables.
- Lifts.

An example of design responsibility for the precast concrete coffered floor slabs is indicated both for the design consultants and the specialist subcontractors.

Precast coffered floor slabs – Design consultants responsibility

- General arrangement drawings and sections.
- Edge condition drawings.
- Loading requirements.
- Finishing specification.
- Setting out.
- Mix design specification.
- Overall performance of floor.

Precast coffered floor slabs – Specialist subcontractor's responsibility

- Mould design.
- Reinforcement design and details.
- Concrete mix design.
- Structural design of precast floors.
- Temporary propping and bracing system.
- Lifting eyes.

## Type B: consultants design with some subcontractor design

The design will be prepared by the design consultant, but elements of design traditionally prepared by specialist subcontractors, such as fixing details etc will be the responsibility of the specialist subcontractor (and so on down to the final design with other packages).

Type B work packages which fall into this category include:

- CFA piling.
- Excavation and earthworks (including retaining wall, basement and ground floor slabs).
- Structural steelwork and decking.
- Concrete frame (staircase cores and structural toppings).
- Secondary steelwork.
- Roller shutter and metal doors.
- Architectural and miscellaneous metalwork.
- Vanity units.

An example of design responsibility for the CFA piling is indicated for both the design consultant and the specialist subcontractor.

CFA piling – Design consultant responsibility

- Piling general arrangement layout.
- Piling loads and piling cut off schedule.
- Piling setting out.

CFA piling – Specialist subcontractor responsibility

- Pile design to engineers loads.
- Pile size and length.
- Detailed design calculations.
- Pile reinforcement design.

With specialist subcontractor packages, the consultant designs are responsible for the general arrangement drawings and layout plans. The package subcontractor is then responsible for the detail design and providing appropriate design calculations. A further example of this is apparent in the concrete frame and cores package (which includes the slipform staircase towers, undertaken by P C Harrington).

Type B work package – Concrete Frame (cores and structural toppings) – Design contractors responsibility

- General arrangement and sections.
- Loading requirements.
- Finishing specification.
- Penetration setting out.
- Mix design specification.

Concrete frame – Specialist subcontractor's responsibility

- Temporary works.
- Slipform system.
- Reinforcement design and scheduling.
- Mix design.

## 10.10  Managing the site logistics

The role of the logistics manager is clearly defined on the Wikipedia website www. wikipedia.org/ wiki/logistics-management

Definitions include:

- Logistics is the management of the flow of resources between the point of origin and the delivery and fixing on site.
- Logistics is the process of planning, implementing and controlling the effective and efficient flow of goods and services from the point of origin to the point of consumption.

## Logistic applications to the project

The logistics manager is part of the site management structure as indicated in section 10.6. On a "must win priority tender" he may be involved in the contract from tender to completion stage.

At the pre-contract stage he will be directly involved with the site layout planning. This will include the preparation of the information shown on the following displays:

• Site layout proposals, location of tower cranes.
• Fire and traffic management plans.
• Logistics approach to concrete pumping to the 12 storey building.

Unloading and lay-down areas for components such as steelwork deliveries, precast concrete floor units and external glazed panels will have to be established. Phased deliveries will need to be planned with manufacturers to meet the site programme requirements for each of the package subcontractors.

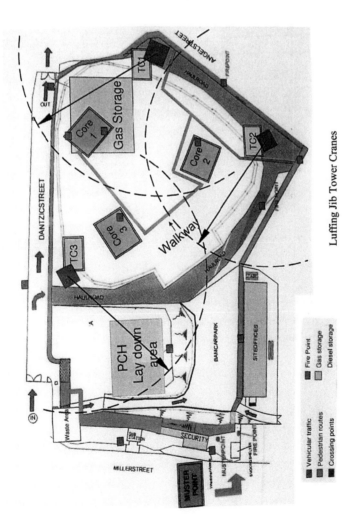

Site Layout showing the location of tower cranes

**Tower cranes on site**

## Fire and traffic management plan

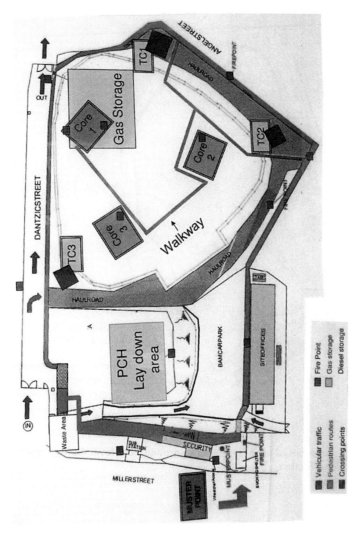

Fire and traffic management plan

## Logistic approach to concrete pumping to 12 storey building

Two static pumps located on access road. Pumps feeding mast pump located in centre of the building.

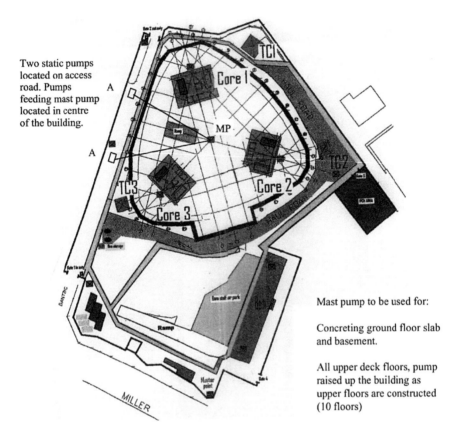

Mast pump to be used for:

Concreting ground floor slab and basement.

All upper deck floors, pump raised up the building as upper floors are constructed (10 floors)

**Schwing concrete pump raised from basement to ground floor level during foundations and ground floor slab construction**

**A: Position of static concrete pump unit on access road**

**Position of mast pump in centre of project**

## 10.11 Sequence of erecting the steel frame and precast floors

The construction sequences is illustrated as a series of diagrams accompanied with site images recorded at the time.

### Sequence one:

Erect precast external columns and steelwork to 2nd floor level. Position precast coffered floor units between steel floor beams to form 2nd floor level.

### Sequence two:

Erect steelwork and precast coffered floor units for floors 3, 4 and 5.

### Sequence three:

Erect steelwork and precast coffered floor units for floors 6, 7 and 8.

### Sequence of erection of the 15 storey steel frame

### Sequence One: ground floor to 2nd floor

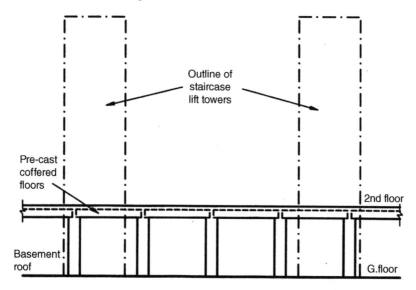

**Erection of pre-cast feature columns from ground floor to second floor**

- Floor Steelwork tied to staircase towers.
- Pre-cast coffered floor units positioned at second floor level.
- See separate details of floor units.

**Pre-cast columns on building façade and 2<sup>nd</sup> floor steelwork and floors completed**

Precast coffered floor units positioned at 2<sup>nd</sup> floor and subsequent floor levels.

**Services provision along underside of floor units**

**Second floor coffered precast units**

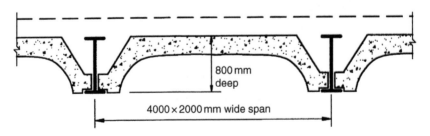

**Precast coffered floor units**

**Handling pre-cast floor units**

## Sequence Two: erect steelwork and floors to 3$^{rd}$, 4$^{th}$ and 5$^{th}$ level

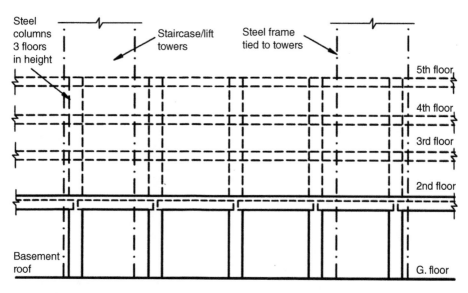

**Erection of steelwork and floors**

**Image of building with steelwork at 5th floor level**

**Main steelwork fixed to staircase and lift cores to provide rigidity to building frame**

**Precast coffered floor units positioned with tower crane working up from floor 3 to 5**

**As the precast floors are completed work can commence to internal walls and service installation on the lower floors**

**General view of floor steelwork**

Position pre-cast coffered units between steelwork to floors 3,4 & 5

## Sequence Three: steelwork and floors 6th, 7th and 8th floor

The steelwork and precast coffered floors were erected over a 6 week period (June/July 2011). Work commenced to the services fit out to the lower ground floors

**Precast coffered floors at levels 6 to 8. Steelwork in progress to elevated level at rear of building (floors 9 & 10)**

## 10.12   Managing the building enclosure

Photo images were recorded of the sequence of work undertaken by the work package subcontractor , Wargner Buro. External glazed wall panels were fixed to the exterior façade of the steel frame.

The proposed method of handling and fixing the panels is as follows:

- Glazed external panels delivered to site and stacked on the storage area adjacent to the lifting gantry.
- At the gantry, the panels are lifted onto a flat bed wheeled trolley. The trolley is moved to the base of the materials lift shaft at ground floor level by a fork lift truck.
- Trolley moved by manual labour into material lift and raised to the selected floor level ie one or two floors above the panel fixing position.

As the trolley is mounted on four air filled rubber tyres it can readily be moved by labourers into position at the edge of the floor slab (note handrail at floor edge).

- External wall panel now lifted off the trolley by small hydraulic crane positioned some three floors above fixing level (see image A on page 191).
- Fixing of brackets at head and base of panel fixed from hydraulic lifting platforms.

A general view of the fixing of the wall panels from floors 3 to 6 is shown. The raising of the panels to floors 7 and 8 was undertaken by a single track monorail trolley traversing the building at floor 8. It was further used for floors 9 to 15 at the rear of the building frame.

## Handling of external curtain wall panels

The handling methods of the external curtain wall panels are as follows:

- Hydraulic tracked mini crane located two to four floors above the fixing level.
- Wall panels on raised trolley platform, raised to appropriate floor level via internal materials lift.
- One fixing gang on hydraulic lifters at bottom edge of panel. Two handlers positioned at head of panel.

**Curtain wall panel in position**

**Image A**

# Handling of external curtain wall panels to floors 3, 4, 5 and 6

**Slings attached to panel and panel lifted from trolley**

**Panel guided clear of building frame**

**Panel lowered into position**

**Panel moved into position and secured in fixing slots**

## Handling of glazed storey height panels by mono rail system to floors 7 and 8

**Wall panels completed to 6th floor using handling method previously shown**

**Fixing of mono-rail support**

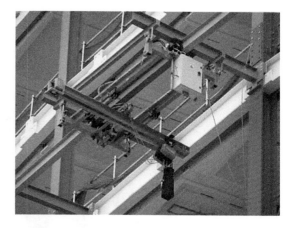

**Single track mono-rail in position for handling glazed storey height panels to rear of building**

## Secondary over cladding to external elevations

The whole of the external elevations of the building are over clad with an independent system of storey height glazing panels. The panels are supported on separate brackets tied back to the main sheet frame.

**Lifting over cladding panels with tower crane**

**Positioning top and bottom of panels from cherry pickers**

**External elevation of building showing secondary cladding**

## External cladding logistics

The project reflects inovative handling and fixing techniques for the exterior facade of a medium rise building. The mixed use of a small hydrolic crane working three floors above the panel fixing level is a well established technique.

The logistics of handling wall panels to the floor 7 and 8 using a single track monorial system is even more innovative.

# Chapter 11

# Chinley School Project

## Contents

| 11.1 | Contents of the case study | 196 |
|---|---|---|
| 11.2 | Project overview | 196 |
| 11.3 | Project information | 197 |
| 11.4 | Client/contractor relationship | 197 |
| 11.5 | Site management structure | 198 |
| 11.6 | Site plan of chinley school | 199 |
| 11.7 | Site layout plan – site compound area | 200 |
| 11.8 | Project manager's involvement at pre-contract stage | 201 |
| 11.9 | Project managers responsibilities during the project | 201 |
| 11.10 | Standard forms and records maintained on site | 202 |
| 11.11 | Materials management on site | 209 |
| 11.12 | Sequence of work and programme – main roof area | 213 |
| 11.13 | Extracts from contract programme | 218 |
| 11.14 | Completed project | 221 |

**Elevation of the completed building**

*Management of Construction Projects*, First Edition. Brian Cooke.
© 2015 John Wiley & Sons, Ltd. Published 2015 by John Wiley & Sons, Ltd.

## 11.1 Contents of the case study

This case study indicates the approach taken to a project under £5 million by a major construction group. The following aspects of the project are highlighted:

- An overview of the project indicating the phasing.
- An outline of the links between the design team, the school co-ordination team and the contractor.
- The approach taken in managing the smaller project by a regional organisation.
- The project manager's involvement at the pre-contract stage and responsibilities during the project.
- The standard company procedures utilised on site.
- An overview of the site waste management plan and examples of materials management practice on site.
- The sequence of work and programme for the main roof work.
- Extracts from contract programmes and the procurement programme.

## 11.2 Project overview

The project involves the construction of a new single residential block to accommodate special needs children.

The work is to be undertaken in four major stages as the existing school premises and access facilities have to remain functional during the building programme.

The building work has been divided into four phases:

1. Phase I: Playing fields and residential block.
2. Phase II: Car park and entrance works.
3. Phase III: New garden centre area.
4. Phase IV Refurbishment of an existing building.

The main monetary value of the project is contained in the single storey residential block (approximately £2 million). Sectional completion dates are applicable to the various phases of the project.

The contract is based on the JCT Standard Form of Contract (2005) with quantities. Two contractors were shortlisted, the contract being awarded on the basis of a negotiated prize and programme.

Interim valuations are monthly between the contractors visiting quantity surveyor and the Derbyshire County Council surveyor.

The contract has been awarded to a national contractor with up to eight regional offices throughout the county. The contract is being managed by the specialist works division undertaking projects up to a value of £5 million.

**General view of completed residential block**

## 11.3   Project information

| | |
|---|---|
| Client: | *Derbyshire County Council Schools Division* |
| Contractor: | *Mansell Construction Services (part of the Balfour Beattie Group)* |
| Contract Value: | *£3.7million* |
| Contract Period: | *13 months* |
| Form of Contract: | *JCT Standard Form (2005) Contract* |
| Start date: | *October 2010* |
| Completion date: | *January 2012* |

The main contractor is required to manage a range of domestic and named subcontractors.

## 11.4   Client/contractor relationship

The following diagram indicates the personnel involved in the project on behalf of the client (Derbyshire County Council). As the current school is in use, it is necessary for the contractor to liaise closely with the school co-ordination team.

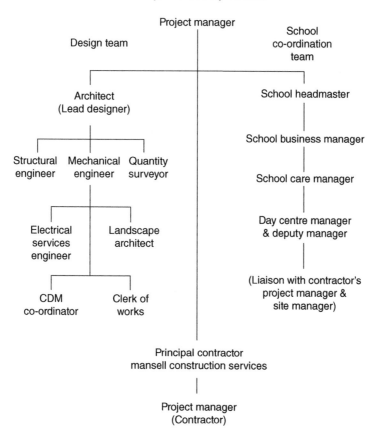

**The structure of the relationships of the client and contractor**

## 11.5 Site management structure

An extensive number of contractors with regional offices, manage contracts under £5 million as illustrated.

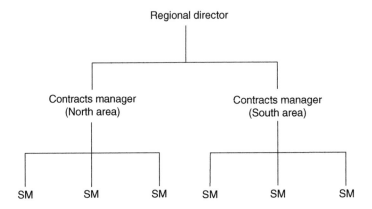

**Example of the management of contracts under £5 million**

Specific management services such as planning and programming, surveying, setting out etc. are undertaken by visiting personnel. These procedures reflect the way in which the Chinley School project is managed.

Management on site is the responsibility of the Construction Project Manager assisted by an assistant site manager.

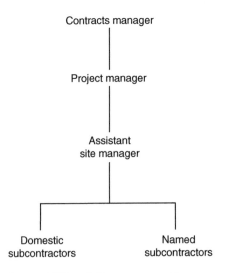

**Responsibility of site management**

The project managers involvement at the pre-contract and contract stage of the contract is later outlined.

The company are keen to develop a paperless office situation on the project. To achieve this objective, the company provide every project manager with a high powered laptop computer containing the complete company procedures i.e. tender, pre-contract and contract documentation. All standard company pro-formas, meeting agendas, weekly and monthly reports etc are included.

## 11.6   Site plan of chinley school

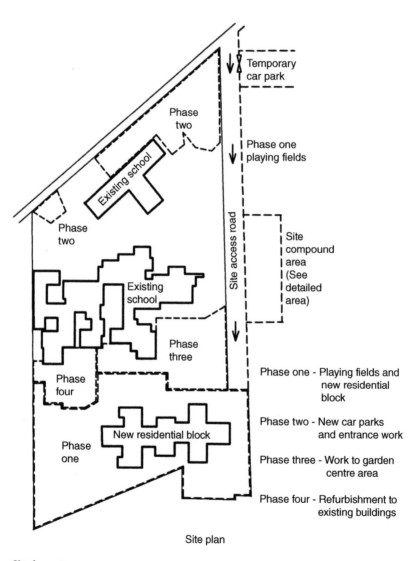

Site plan

**Site layout**

## 11.7 Site layout plan – site compound area

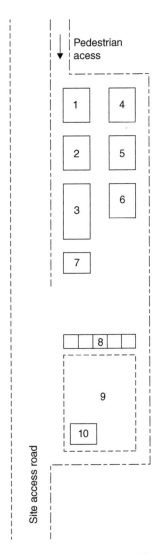

**Good site layout plan with organised compound and storage areas**

**Site compound area**

1. Site office - contractor
2. Mess and drying room
3. Subcontractor's stores
4. Meeting room and clerks of works
5. Toilets – male and female
6. Store
7. Outside smoking area
8. Waste management containers
9. Materials compound
10. Diesel store

**Materials compound**

## 11.8 Project manager's involvement at pre-contract stage

Many large companies develop project teams for undertaking both medium and larger projects. It is common practice to appoint the project manager during the pre-contract period, some 6-8 weeks prior to the commencement of work on site.

During the pre-contract period the project manager may be involved in:

* Preparing the contract programme with the appointed planning engineer. Familiarity with the programming software is an essential requirement of this process. In this case the manager should be capable of using Power Project (Asta Developments Software package).
* Preparing schedules of information requirements based on the contract programme.
* Preparing the short term planning for the first 4 to 6 weeks of the project.
* Preparing the site layout plan, access roads and material storage areas.
* Organising the movement of office accommodation and plant for the initial work stages.
* The preparation of a site safety plan and method statements for the initial operations on site.
* Ensuring all statutory notices have been served ie water supply, electrical supply to cabins and telephones.
* Agreeing a workable cost budget for labour and plant.
* Attending the pre-contract meeting with the client, design team and contractors personnel involved in the project.

## 11.9 Project managers responsibilities during the project

* To act as the contractor's representative with respect to the client. All communications should be directed through the project manager.
* To manage the project on a daily basis with respect to all site based personnel. Certain responsibilities may be delegated to assistant site managers and site supervisory personnel.
* To maintain daily records. This includes:
    Daily site diary.
    Record of maximum and minimum temperatures.
    Visitor records.
    Labour, plant and materials movement.
* To ensure implementation of the site safety plan and safety records. All safety registers are to be maintained with regard to site inspections of excavations and scaffolds etc.
* To manage all site meetings with regards to the client and subcontractor's. This includes producing progress records and proposals to reduce delays to the contract programme.
* To overview the project cost and value position with the project surveyors on a monthly basis.
* To maintain the safety file during the project for handover to the client on completion.
* The monitoring of progress during the project is an essential part of the managers responsibility. Assistance may be provided by a visiting site planning engineer responsible for updating and preparing accelerated programme proposals.

- Managing the inspection and handover process with the subcontractors.
- To prepare monthly reports on site progress, highlighting project delays. Proposed measures for accelerating sections of work which are behind programme.
- To organise monthly site meetings with the architect and subcontractors representatives.

## 11.10 Standard forms and records maintained on site

One of the main contractor's objectives is to produce a "paperless office system". To aid this process, a wide range of standard procedures have been provided via the company's web site. This can be accessed by the project manager on a laptop computer.

The range of standard forms include:

1. Weather and temperature record sheet.
2. Record of visitors to site (daily).
3. Confirmation of verbal instructions.
4. Weekly site report.
5. Monthly progress report.
6. Subcontract production meeting.

Brief notes are indicated on the purpose of each of the above forms.

**The residential block- blockwork and brickwork to first lift**

## 1. Weather and temperature record

| WEATHER and TEMPERATURE RECORD | | | | | |
|---|---|---|---|---|---|
| Job No. | | | | Job Title | |
| Date | 08.00 | 10.00 | 12.00 | 14.00 | Notes and weather |
| | | | | | |
| | | | | | |
| | | | | | |
| | | | | | |
| | | | | | |

Example of a weather and temperature record form

Weather and temperature records are necessary in order to record days lost due to exceptional inclement weather (i.e. extreme temperatures which effect concreting operations and heavy rainfall which results in the stopping of work).

## 2. Record of visitors to site (daily)

| SITE VISITORS | | | Job Title | |
|---|---|---|---|---|
| Job No. | | | Date: | |
| Visitors | Time In | Time Out | Purpose of visit | Signature |
| | | | | |
| | | | | |
| | | | | |
| | | | | |
| | | | | |

Example of a record of visitors to site form

All site visitors are required to sign in and out at the gatehouse or reception. Before entering the work areas, site induction is necessary and the requirements for P.P.E. must be complied with i.e. safety helmet, protective eye wear, gloves, safety boots and protective vests.

## 3. Confirmation of verbal instructions

---

### CONFIRMATION OF VERBAL INSTRUCTION

**Sheet No.**    **Job No.**    **Date:**

**Job Title**

**Instructions Received From:** *Architect and CoW*    **of:** *Cox and Partners*

**INSTRUCTION:**

*We confirm the following in terms of works to the Garden Centre:*

*Installation of a timber fillet to the shed roofs:*
*Installation of gutters and rainwater pipe to 2 no. sheds.*
*Installation of concrete strip sloping between the shed bases for maintenance.*
*Installation of lead slate for the soil and vent pipe to the kiosk.*

**To: Contract Administrator:**

**WE ACKNOWLEDGE HAVING RECEIVED THE ABOVE INSTRUCTIONS/DRAWINGS AND THIS MATTER WILL RECEIVE OUR PROMPT ATTENTION**

**IF THIS INSTRUCTION IS NOT DISPUTED WITHIN SEVEN DAYS, THEN THIS WORK WILL BE COMPLETED AND CHARGED ACCORDINGLY, EITHER ON A MEASURED BASIS OR DAYWORK BASIS AS APPLICABLE**

**Signature:**    **Name:**

---

**Example of a confirmation of verbal instructions form**

*Notes from Contract Practice:* Most of the organised construction firms use CVIs (confirmation of verbal instruction). These may be in a simple pad format. The contract contains clauses relating to the procedure for confirming verbal instructions either from the architect or the clerk of works. The confirmation of architects or other verbal instructions given by the consulting engineer, may be recorded in the monthly site minutes. A separate heading should be included in the minutes to ensure that variations are raised at the appropriate stage in the meeting agenda.

## 4. Weekly site report

On the small/ medium sized projects, the weekly site report will be compiled by the site manager/ project manager. The main purpose is to report on the present and future labour and subcontract requirements.

A review of the progress position is indicted together with an assessment of further delays likely to occur. A summary of key information requirements is highlighted (with respect to labour, plant, materials and subcontractors). Outstanding site instructions may be shown together with key drawings awaiting issue.

| WEEKLY SITE REPORT | | Week No.: | | | Date of report: | | | |
|---|---|---|---|---|---|---|---|---|

| Job No.: | Job Title: | | | Site Manager: | | | | |
|---|---|---|---|---|---|---|---|---|

**Period of Delay Notified:**     **Extension of Time Granted:**

| | Sat | Sun | Mon | Tue | Wed | Thur | Fri |
|---|---|---|---|---|---|---|---|
| Weather Comments and Effects on Progress | | | | | | | |

| Labour on Site | S | S | M | T | W | T | F | Required Wk1 | Wk2 | Wk3 | Wk4 | S/c requirement within 4 weeks for which no arrangements concluded or Order received |
|---|---|---|---|---|---|---|---|---|---|---|---|---|
| | | | | | | | | | | | | |
| | | | | | | | | | | | | |
| | | | | | | | | | | | | |
| | | | | | | | | | | | | |
| | | | | | | | | | | | | |
| | | | | | | | | | | | | |
| | | | | | | | | | | | | |
| | | | | | | | | | | | | |
| | | | | | | | | | | | | |
| | | | | | | | | | | | | |

**Programme: List foreseen future delays**

| Drawings and all Information Outstanding: | Site Instructions Not Confirmed/Drawings and Information Issued Direct to Site: |
|---|---|

| Visitors: Including all Contractor, Subcontractor and Consultant Personnel | Distribution |
|---|---|
| Sat | |
| Sun | |
| Mon | |
| Tue | |
| Wed | |
| Thur | |
| Fri | |

Distribution: Office, Site

**Example of a weekly site report form**

## 5. Monthly progress report

It is an unusual procedure to plan the monthly progress meeting just prior to the monthly site meeting with the client. This allows progress on each work section to be discussed with the subcontractors and supervisors. This ensures that at the clients monthly site meeting, everyone "sings from the same hymn sheet".

A review of site safety, quality control and the waste management situation may be reviewed. A review of the progress of each subcontractor's package may be agreed and delays realistically assessed.

| MONTHLY PROGRESS REPORT | |
|---|---|
| **Job No.:** Job Title: | |
| **Job No.:** 28   Programme No.: (20-08-10)   Report No.: 07 | |
| **Prepared On:** 25-03-11 **Prepared By:** A.P.Hodgkinson   **Approved By:** D.Williams | |
| **Summary Information** | |
| Section 1 - Completion date - 24-10-11 | Anticipated Completion Date: 23-12-11 |
| Section 2 - Completion date - 04-01-11 | *Anticipated Completion Date: 08-04-11* |
| Section 3 - Completion date - 28-02-11 | Anticipated Completion Date: 10-06-11 |
| Section 4 - Completion date - 25-07-11 | Anticipated Completion Date: 05-09-11 |
| Section 5 - Completion date - 30-08-11 | Anticipated Completion Date: 23-12-11 |
| *Late details for the kiosk M&E installation has meant late procurement, ceiling cannot be installed until electrical wiring has been undertaken. A proposed date for completion of the works is 22nd April 2011. This does not include the Mains electric supply, metering and final test and commissioning.* | |
| *Please refer to section drawing for completion dates for areas of wotks (previously issued)* | |
| **Safety, Quality, Environment and Security Summary** | |
| **Safety** *No accidents have been recorded on site over the reporting period. The Project Management Plan (Construction Health & Safety File) needs to be updated to include Brian Cooke as the Site Manager and his contact details. This will be issued in due course. All operatives have been inducted and are working under an approved method statement and risk assessment. The Work Activity Schedule is up to date with the above information. Tool box talks have been issued to all site operatives as per company procedures.* *Man-hours worked on site without an accident, upto 25-03-11 was 6101 hrs* *Over reporting period 25-02-11 = 5043 hrs* *To date from 13-09-10 to 25-03-11 20168 hrs* **Quality** *Numerous visits undertaken from CoW, inspecting the external masonry. Issues raised and addressed with the stonework and damp details which has been addressed. Inspection of roof trusses undertaken 25-03-11, issues raised with clips (refer to TQ raised)* **Environment** *Site Waste Management Plan is in place and is held on site. The Hazardous Waste Registration is: NVW396* **Security** *No issues to date.* | |
| **Building Control** | |
| *Inspections undertaken on drainage and masonry works throughout the site.* | |

**Example of a monthly progress report form**

## 6. Subcontractors production meeting

| SUBCONTRACTOR PRODUCTION MEETING | | Page 1 of 2 |
|---|---|---|
| **Job No.:** | **Job Title:** *Ormskirk School, Ormskirk, Lancashire* | |
| **Trade(s):** | **Subcontractor(s):** *Northern Rock Asphalt* | |
| **Present:** | | |
| **Date:**    **Last Period:**    **To:**    **Next Period:**    **To:** | | |
| **Start Date:**    **Overall Duration (Wks):**    **Current Programme:** | | |

| Agenda | Minutes | Action |
|---|---|---|
| **Previous Meeting** | | |
| Actions | *First meeting* | |
| **Safety Performance Review** | | |
| Accidents/Incidents | *None* | |
| Inspections and Audits (By Contractor and Subcontractor) | *None* | |
| Inspections and Audits (By Contractor and Subcontractor) | *None as yet, due this week* | |
| **Safety Planning** | | |
| Work Activities (Does the Work Activity Schedule confirm imminent Significant risk?) | *Yes* | |
| Control Arrangements (Has the Subcontractor provided suitable risk assessments | *Up to date* | |
| Inspections (Dates and resources) | *Crash deck in place, but to be moved as work progesses* | |
| Training (Briefings and resources) | *As above for toolbox, NRA to send training matrix* | |
| **Quality Performance** | | |
| Samples/Mock-ups | *As per specification* | |
| Inspections & Defects | *CoW inspections recorded fixing of 20mm insulation Inspections all up to date* | |
| Corrective Action | *NTR* | |
| **Environmental Performance** | | |
| Management of Waste | *Monitored, skips provided & being used* | |
| Management of Nuisance | *Excess waste of insulation due to cuts up hips & valleys* | |
| **Labour** | | |
| Levels | *5 No. at present, due to be monitored* | |
| Supervision | *1 No.* | |

**Example of a subcontractor production meeting report form page 1**

| SUBCONTRACTOR PRODUCTION MEETING | | Page 2 of 2 |
|---|---|---|
| **Agenda** | **Minutes** | **Action** |
| **Materials and Plant** | | |
| Status of Supply | *On site and progressive deliveries* | |
| Waste | *To be Monitored* | |
| Deliveries | *No issue* | |
| **Information Required** | | |
| Subcontractor Information Required Schedule | *Up to date* | |
| CVI's/TQ's | *As specification, up to date* | |
| Subcontractor Drawing Issue Status | *Ok* | |
| Technical Submissions | *Approved issued for timber fixings & monaperm membrane* <br> *Flats on internal of truss to hold insulation* | *M6* |
| Procurement Schedule | | |
| **Design** | | |
| Drawings | *N/A* | |
| Approvals | *N/A* | |
| Costs | *N/A* | |
| **Co-ordination** | | |
| Other subcontractors | *Rooflights to be constructed (setting out)* <br> *Facia installed by cover over brick & cavity to be* <br> *Shop to be installed* <br> *Lighting protection to be coordinated* | |
| **Commercial Issues** | | |
| Variations | *Floor timber instructed for kiosk* | |
| Dayworks | | |
| Work Completed by Others | *As above for S/C* | |

**Progress/Short Term Programme**

| Operation | % Done During last period | Total % Now Complete | Programme % Progress To Date | % Target during Next Period |
|---|---|---|---|---|
| *Insulation & Membrane progress to west end of building. Working in line with current requirements.* | *Completed By* | | | |
| *ZONE 1 - INSULATION & LATT* | *21/4/11* | *Half Bld* | | |
| *ZONE 2 - INSULATION & LATT* | *6/5/11* | *Completed* | | |
| *ZONE 3 - INSULATION & LATT* | *13/5/11* | *By 28/4/11* | | |
| *Decision made to get water tight ASAP* | | | | |

**Progress/Programme Comments**

| |
|---|
| |

**Next Meeting Date:** *10/5/11*

**Example of a subcontractor production meeting report form page 2**

## 11.11   Materials management on site

### The site waste management plan

The following is an overview of the purpose of headings identified on the waste management system. The displays are used to identify:

- The operations creating the waste. This is an operational approach to the waste created from the demolition stage through to the final internal and external operations. Each stage of work creates its own waste problems.
- Types of waste streams generated i.e. active waste, inert waste and hazardous waste.
- The quantities of waste (tonnage) in each of the waste streams.
- The strategy for waste disposal of the waste created.

The European waste category code should be shown together with details of the waste carrier responsible for the disposal. The relevant licence numbers, consignment number and note number may be displayed.

The form also includes a summary of:

| | |
|---|---|
| Total waste (tonnage) | 2911 T |
| Re-used/recycled on site | 92 T |
| Sent off site for recycling | 942 T |
| Sent to authorised disposal | 1830 T |
| Sent to land fill | 47 T |

The objective of waste management is to dispose the minimum quantity generated to landfill. A target of 5% maximum for material deposited in landfill is aimed at by the company.

## The site waste management plan

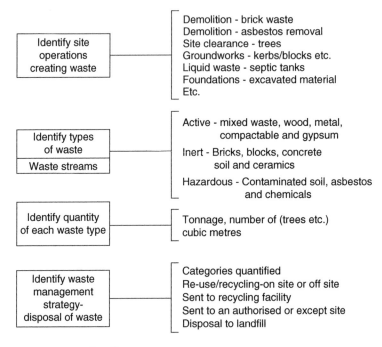

| | |
|---|---|
| Identify site operations creating waste | Demolition - brick waste<br>Demolition - asbestos removal<br>Site clearance - trees<br>Groundworks - kerbs/blocks etc.<br>Liquid waste - septic tanks<br>Foundations - excavated material<br>Etc. |
| Identify types of waste<br>Waste streams | Active - mixed waste, wood, metal, compactable and gypsum<br>Inert - Bricks, blocks, concrete soil and ceramics<br>Hazardous - Contaminated soil, asbestos and chemicals |
| Identify quantity of each waste type | Tonnage, number of (trees etc.) cubic metres |
| Identify waste management strategy-disposal of waste | Categories quantified<br>Re-use/recycling-on site or off site<br>Sent to recycling facility<br>Sent to an authorised or except site<br>Disposal to landfill |

**Site Waste Management Plan diagram**

**Site Waste Management**

| Summary of Waste Management Plan | | | | | | | | |
|---|---|---|---|---|---|---|---|---|
| Activity | Waste Stream | | | Waste Management Strategy | | | | |
| Material Produced | Active | Inert | Haz. | Re-use | Recycle | Auth. Site | Landfill | Carrier |
| Demolition - bricks | | ✓ | | 20 | | | | Use as hardcore |
| Site clearance - topsoil | | ✓ | | 4545 | | | | Reuse on site |
| Groundworks - kerbs | | ✓ | | | 15 | | | Recycle facility |
| Septic tank waste | | | ✓ | | | | 40 | Licenced tip |
| Internal plasterboard | ✓ | | | 35 | | | | Spec. recycling |
| Plumbers, B/L, dryliner | ✓ | | | 10 | | | | Recycling |
| Hazardous waste bins | | | ✓ | | | 2 | | Recycling |
| Site office paper | ✓ | | | | 1 | | | Recycling |
| Canteen/cans, plastic | ✓ | | | | 1 | | | Recycling |

**Summary of Waste Management Plan**

## Materials management on the project

**Materials management on site**

Good practice procedures are evident on the project with extensive use of timber pallets to ensure that the materials are stacked clear of the ground. Subcontractors are encouraged to place loose material on pallets adjacent to the works. Timber off-cuts are to be placed in mini-skips. Materials stored on pallets are placed on pre-prepared stoned up areas.

**Materials stored on pre-prepared stoned up areas**

## Good practice observations

**Palleted goods, banded and stored adjacent to access roads**

**Stone lintels stored for ease of handling**     **Plastic goods delivered in easy hand able sacks**

**Use of dedicated skips for selective materials such as timber, insulation off cuts etc.**

## 11.12 Sequence of work and programme – main roof area

A large span timber trussed roof covers the whole of the new residential block. The roof plan is indicated showing the layout of the ridges, hips and valleys. Work to the roof area is to be undertaken in three phase's i.e.

- Area One – South roof and gables
- Area Two – Central roof and gables
- Area three – North roof and gales

On completion of the fixing of trusses to Area Two the roof coverings to Area One is to be commenced.

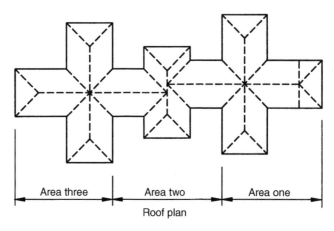

**Diagram of the roof plan**

The sequence of work in each area involves the following operations:

- Position Safety Deck (scaffolding operation)
- Fix wall plates
- Unload truss and fix trusses and bracing
- Work to facia and roof vents
- Roof finishes (tiling) – subcontractor.

## Roof to residential block

Sequence of work:

**Safety Deck system in position 900 mm below the underside of the trusses**

**Alternative Safety Deck system**

**225 mm x 75mm Softwood timber wall plates bolted to steelwork or fixed to top of brickwork**

- Position Safety Deck on tubular support under roof trusses in Area One to allow access for fixing.
- Fix timber wall plates to external wall perimeter.
- Delivery and unloading of 10 m span roof trusses.
- Position trusses on roof and temporary strap in vertical position.

**Unloading and lifting of 10 m span trusses into position**

**Trusses positioned in Area One**

## Main roof – site progress images

### Roofing work to area one

**Work in progress in Area One. Progress after 5 days of work**

**Internal bracing to 10 m span roof trusses. Safety Deck shown for access to underside of trusses**

## Main roof – site progress images

### Roofing work to area three

**Completion of structural timber work and roof vents**

**120 mm insulation to roof area prior to felting and battening**

**Slating to main roof in progress**

## Main roof construction

### Safety deck system

The works at Height Regulations 2005 are applicable to all work at heights where there is a risk of falls liable to cause personal injury. HSE statistics indicate that falls from height are the most common causes of fatal injuries and the second most common cause of injury to employees.

Safety Deck is a passive system that can be used to prevent a fall. The system is cost effective and is used by major contractors (ref. website safetydeck@tarmac. co.uk ). The system allows full freedom of movement during the installation of roof trusses, timber and pre-cast flooring systems and flat roofs. The Safety Deck website indicates the full benefits of the system.

## Programme for the roof construction

An extract from the contract programme is shown for the roof construction and coverings. The overall period is shown as 13 weeks including the solar panel installation. Progress has been recorded to the end of week 28. The programming is based on linked bar-charts using Asta Developments Power Project Software.

### Acceleration of roofing programme

Due to extensive inclement weather in December the contract is currently some 8 weeks behind programme. The project manager has considered that the roof programmes may be accelerated in order to recover part of the delay. Operation 3 (erect roof trusses and roof carcase) and Operation 6 (roof coverings) may be accelerated. A detailed short term programme for roof trusses and roof coverings is illustrated, showing a reduction of 5 days. Further reductions may be considered by increasing the number of subcontractor gangs fixing the roof coverings.

| Line | Name | Planned percent complete | Percent complete |
|------|------|------|------|
| 1 | Top out walls, bed & lay wall plates in prep for trusses | 50% | 50% |
| 2 | Fall arest protection | 50% | 50% |
| 3 | Erect roof trusses, roof carcass & BWIC solar panel | 23% | 30% |
| 4 | Insulation, battens & membranes | | |
| 5 | Lightning protection | | |
| 6 | Roof coverings | | |
| 7 | Rooflight installations | | |
| 8 | Rainwater goods & guttering: including connections | | |
| 9 | Solar panel installation | | |

**Extracts from contract programmes**

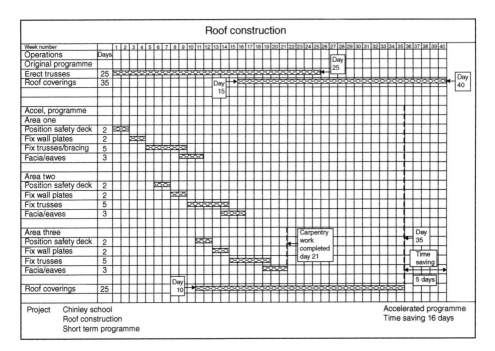

**Bar chart – roof construction to Areas 1, 2 and 3**

## 11.13 Extracts from contract programme

The company use standardised planning procedures on all major projects. The planning department support all planning stages from tender to project completion.

Planning engineers may be site based on larger projects or simply visit the site at regular weekly/monthly intervals to upgrade the contract programme. Programmes will initially be upgraded prior to monthly site meetings and a progress report prepared with the site manager.

The company policy is to present all programmes using linked bar-chart principles based on the Asta Development Software/Power Project.

The following programmes have been illustrated for the Chinley School Project:

• Extract from retaining wall programme.
• Extract from internal finishes programme.
• Extract from procurement programme.

The software used for project planning is Asta Developments – Power Project / Team Plan suite of programmes. The managers are familiar with the use and application of Power Project and are capable of producing site short term programmes for any stage of work.

## Master programme extract

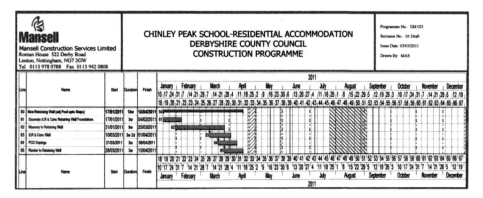

**Extract from site programme for the retaining wall section**

**Extract from site programme for internal services and finishes**

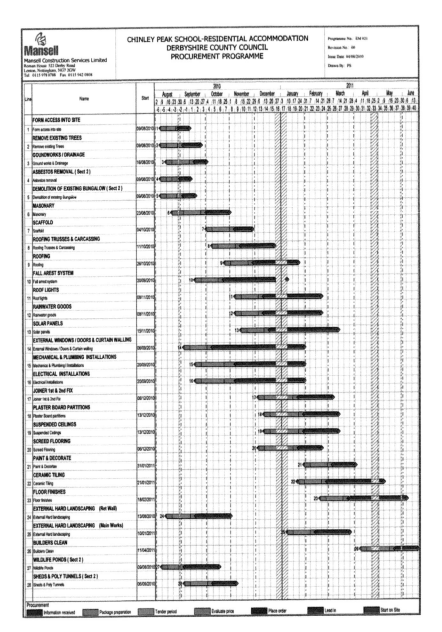

**Procurement programme**

## 11.14   Completed project

### New residential block

The completed building is a complex new build residential block incorporating extensive landscaping work.

Value:       *£3.7 million*
Period:      *13 months*
Contractor: *Mansell Construction Services*
Client:       *Derbyshire County Council*

**Completed building**

# Chapter 12

# Retail Unit and Car Park

## Contents

| | | |
|---|---|---|
| 12.1 | Contents of the case study | 223 |
| 12.2 | Project overview | 224 |
| 12.3 | Project information | 224 |
| 12.4 | Client/contractor relationship | 224 |
| 12.5 | Project management structure | 226 |
| 12.6 | Site management structure | 227 |
| 12.7 | Site layout plan | 228 |
| 12.8 | Construction stages – sequence of work glulam frame to main store | 231 |
| 12.9 | Risk assessment – concrete operation | 246 |
| 12.10 | Concrete placing to car park slab | 246 |

*Management of Construction Projects*, First Edition. Brian Cooke.
© 2015 John Wiley & Sons, Ltd. Published 2015 by John Wiley & Sons, Ltd.

## 12.1  Contents of the case study

**Elevation of the completed building**

This case study covers the construction and management aspects in building a timber portal framed building. The £11 million project was completed over a relatively short ten month construction period.

The following aspects of the project are covered in this case study:

- The client/contractor relationships are outlined together with the contractors approach to managing the project.
- The project management structure for the main contractor highlighting the link between senior management and site management.
- The site layout plan indicating the relationship between the main buildings under construction and the storage areas on site.
- The construction stages for the erection of the glulam timber frame relating to the foundation stages and the erection sequence for the nine bays forming the retail store. Interesting connections are shown at the base and roof level.
- An interesting method statement is shown for the concreting of the first floor of the car park building. Risk assessment areas are identified for concreting operations and pouring sequences are shown for the complete floor pour.

## 12.2   Project overview

The project involves the construction of a supermarket building on the outskirts of Manchester. The building incorporates a timber framed two storey height structure forming the retail sales area. Adjacent to the building is a two storey steel framed car park adjoining the building (see site layout plan) and a vehicular distribution centre is located at the rear of the retail unit.

A further phase of the development is to include a community centre and a petrol station.

## 12.3   Project information

Client/Developer:   *CTP Ltd*
Main Contractor:   *Morgan Sindall*
Contract Value:   *£11 million*
Contract Period:   *10 months (42 weeks)*
Completion date:   *June 2012*
Form of Contract:   *JCT 2005 Design and Build*
*Single Stage Tender*
*Novated Design Team*

Major work packages:

- Foundations and earthworks
- Glulam timber frame – Bowmer & Kirkland Structures
- Steelwork – Retail store mezzanine floor and car park frame
- Insitu concrete floors
- External SIP panels
- Roof decks
- External glazed walls
- Fit out services
- Heating/lighting and ventilation services

## 12.4   Client/contractor relationship

The client (CTP Ltd.) is the developer who is providing the retail store and car park on a lease to a major retailer (Tesco) for a period of some years.

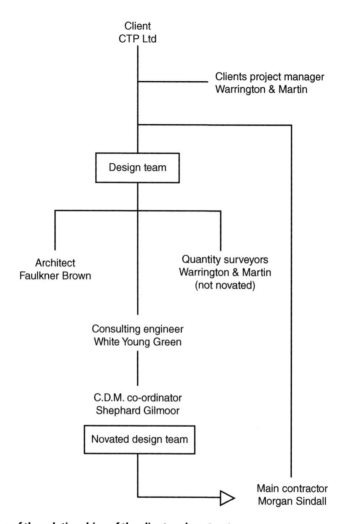

Client
CTP Ltd

Clients project manager
Warrington & Martin

Design team

Architect
Faulkner Brown

Quantity surveyors
Warrington & Martin
(not novated)

Consulting engineer
White Young Green

C.D.M. co-ordinator
Shephard Gilmoor

Novated design team

Main contractor
Morgan Sindall

**The structure of the relationships of the client and contractor**

## 12.5 Project management structure

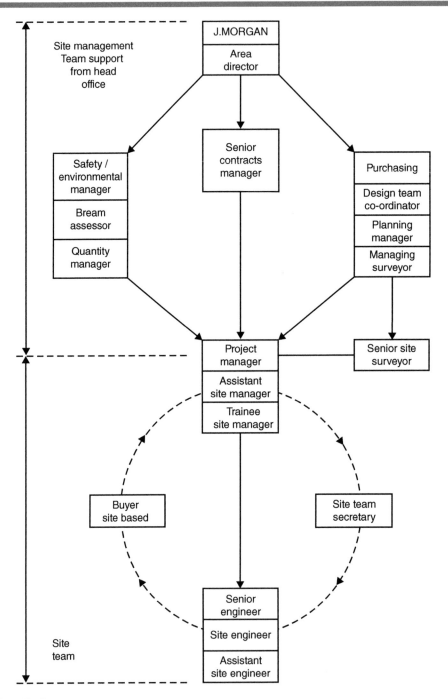

**Tesco Project**

## 12.6 Site management structure

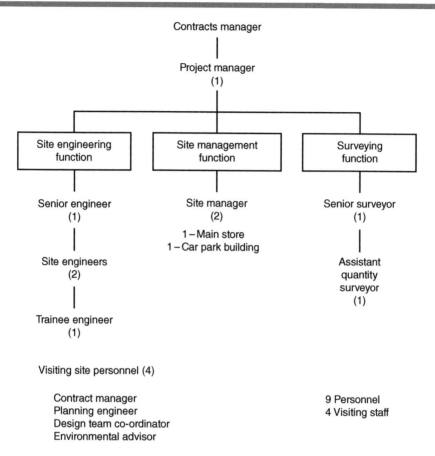

Contracts manager

Project manager
(1)

| Site engineering function | Site management function | Surveying function |

Senior engineer
(1)

Site manager
(2)

Senior surveyor
(1)

1 – Main store
1 – Car park building

Site engineers
(2)

Assistant
quantity
surveyor
(1)

Trainee engineer
(1)

Visiting site personnel (4)

Contract manager                    9 Personnel
Planning engineer                   4 Visiting staff
Design team co-ordinator
Environmental advisor

**Structure of the site management**

## 12.7 Site layout plan

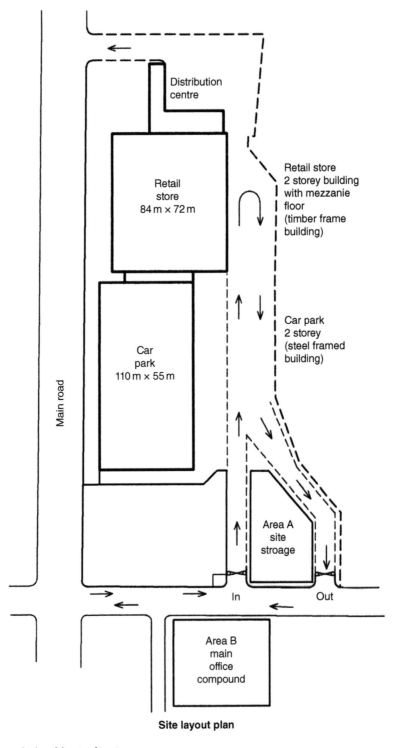

**Site layout plan**

**Building relationships to site storage areas**

# Site layout plan – material storage areas

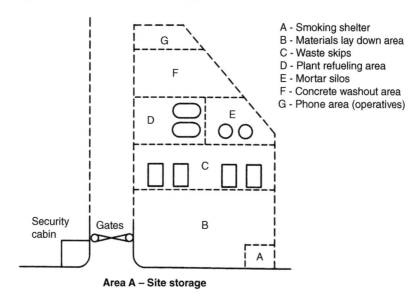

A - Smoking shelter
B - Materials lay down area
C - Waste skips
D - Plant refueling area
E - Mortar silos
F - Concrete washout area
G - Phone area (operatives)

**Area A – Site storage**

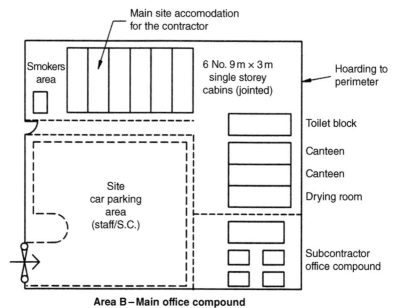

**Area B – Main office compound**

**Area A – Site storage & Area B – Main office compound**

## Main site accommodation for the contractor (Area B)

**Main site office compound**

The main site office compound is located on an adjacent area of land opposite the main site entrance. This forms a main contractor and subcontractor area including a large site car park facility.

**Main entrance foyer to site office**                **Sample panels adjacent to office**

The office and subcontractor facilities are contained in 10 number (9 m x 3 m) single storey linked container cabin units.

## 12.8 Construction stages – sequence of work glulam frame to main store

The laminated timber frame has been designed and erected by a specialist firm as a work package. The 300 x 450mm columns are arranged on a 6m and 18m grid layout. A mezzanine floor is incorporated in the store supported on a steel frame to form a first floor sales area. The ground floor and mezzanine floor are finished with an insitu concrete slab. Externally the frame is clad in large SIP panels two storeys in height. Further large glazed wall panels complete the building enclosure.

**End gable frame showing insulated SIP timber cladding panels**

**High level glazing panels are located around the building**

**10 bays of timber laminated columns at 6m centres**

## Glulam timber frame – foundations

The Glulam timber columns and roof beams were provided as part of the work package for the six metre high frame. This was designed and erected by BK Structures, part of the Bowmer & Kirkland Group based in Derby.

A construction period of 10 weeks was allowed in the tender for the erection of the complete frame (the foundation works were completed as a separate work package).

### Stage 1

Excavate rectangular pad foundations, fix reinforcement, holding down bolts and concrete pad foundations.

### Stage 2

Glulam timber columns delivered to site in plastic shrink wrap protection. Plated steel base and head connections to timber columns to be fixed on site prior to lifting columns and beams into position.

## Stage 3

Glulam columns at 6m centres to form a 60m long building (11 columns per row). On completion of two rows of columns, glulam roof beams are positioned to form 18m wide bays (see layout plan).

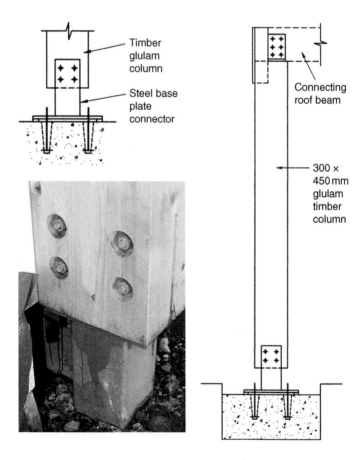

Timber glulam column

Steel base plate connector

Connecting roof beam

300 × 450 mm glulam timber column

## Glulam timber frame – connections to timber frame

**Glulam timber frame**

As stated previously, these connecting plates are fitted to the edge beam and roof beam ends prior to them being lifted into position. A bolted connection is then made to complete the joint assembly from a hydraulic "cherry picker".

**Steel tubular bracing at rook level**

## Glulam timber frame – roof structure

**Internal view of roof structure on 6m x 18m long grid.**

Steel bracing to 18m spans

**Metal base plates and column connections to all laminated columns and roof beams. This enables the speedy erection of the main timber frame**

## Glulam timber frame – bay erection sequence

| 9 | 8 | 7 |
|---|---|---|
| 6 | 5 | 4 |
| 3 | 2 | 1 |

**Bay layout. Erection sequence**

**Bay 1 – frame erection day 5–7 on programme**

**Bay 2/3 – frame erection day 8–10 on programme**

**Bay 4 – frame erection day 14–15 on programme**

## Glulam timber frame – erection sequence

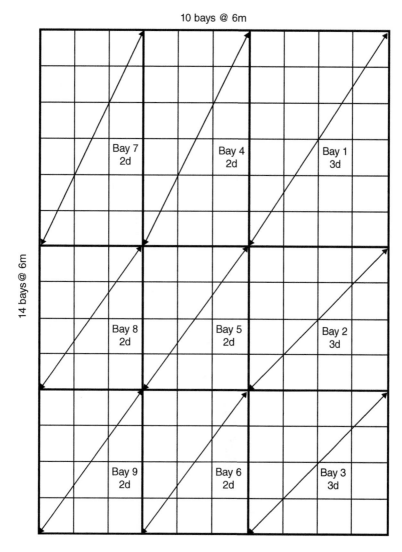

10 bays @ 6m

14 bays @ 6m

Bay 7
2d

Bay 4
2d

Bay 1
3d

Bay 8
2d

Bay 5
2d

Bay 2
3d

Bay 9
2d

Bay 6
2d

Bay 3
3d

Overall duration for glulam frame erection – 21 days
(contract programme allowance 20 days)

**Plan of main store Glulam frame**

# Glulam frame – linked bar chart programme

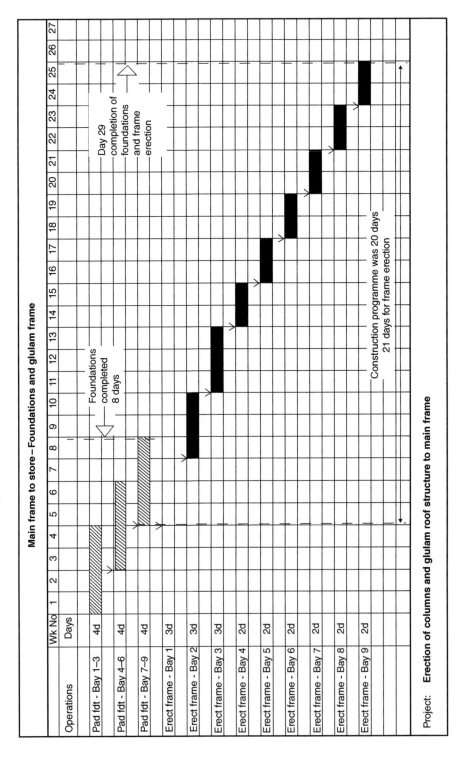

**Main frame to store—Foundations and glulam frame**

| Operations | Days | Wk No | 1 | 2 | 3 | 4 | 5 | 6 | 7 | 8 | 9 | 10 | 11 | 12 | 13 | 14 | 15 | 16 | 17 | 18 | 19 | 20 | 21 | 22 | 23 | 24 | 25 | 26 | 27 |

Pad fdt - Bay 1–3 — 4d

Pad fdt - Bay 4–6 — 4d

Pad fdt - Bay 7–9 — 4d

Erect frame - Bay 1 — 3d

Erect frame - Bay 2 — 3d

Erect frame - Bay 3 — 3d

Erect frame - Bay 4 — 2d

Erect frame - Bay 5 — 2d

Erect frame - Bay 6 — 2d

Erect frame - Bay 7 — 2d

Erect frame - Bay 8 — 2d

Erect frame - Bay 9 — 2d

Foundations completed 8 days

Day 29 completion of foundations and frame erection

Construction programme was 20 days
21 days for frame erection

**Project:** Erection of columns and glulam roof structure to main frame

**Foundation and glulam frame**

## Method statements

The CDM regulations require the contractor to prepare method statements for approval by the clients representatives prior to works being undertaken. Likewise subcontractors are required to submit method statements prior to them undertaking operations on site.

Method statements are directly linked to "safe methods of working" and hence to risk assessments. At the subcontractor's induction meeting on site, emphasis is placed on the fact that operations cannot proceed without familiarity with both the task method statement and the associated risk assessments.

No standard format for presenting method statements has yet been developed. They may be presented in written form or as a tabular presentation. Method statements may be part of the tendering process as many tenders are won on a construction method approach. When considering alternative approaches at the "value engineering" stage of a design and build project a construction method approach is essential.

The areas considered during the preparation of a method statement for concreting operations and associated risk assessments are shown. Extracts from site specific method statements are outlined for concrete laying and finishing to the car park area.

A risk assessment format is shown for fire hazards likely to be encountered.

## Method statement for suspended floor to steel framed building (car park). 150mm Insitu suspended floor to steel framed building (car park)

### Description of Task

The method statement and risk assessment covers the placing of 150 mm insitu concrete to a metal deck floor. The work is to be undertaken by a specialist floor laying contractor as part of a package contract on a Design and Build project. The method statement and risk assessment have been produced for the contractor by a safety consultant.

### Sequence of Work

A plan of the first floor car park is shown. This indicates that the suspended floor is to be poured in six 18m wide bays across the width of the building. The access ramp to the first floor is to be poured separately (bay pour 7). The quantity of concrete per pour approximates some 122 cm. The pouring sequence is based on pouring one bay per day from bay 1 to 7. The stop end between bays is to be removed prior to the next pour.

## Site specific to concreting operations

**Method Statement Headings**

- Emergency Aid Procedures
  Emergency Aid Certificate holders
  Emergency Procedures
- Personal Protective Equipment
  List of PEP available

- Direct and subcontractor labour requirements

- Plant requirements – list of types of fuel used

- Traffic management proposals

- Work area protection

- Material deliveries
- Set up tasks prior to concreting
  Establishing dedicated unloading, access
  areas and prior tasks such as fixing fabric
  reinforcement

- Access

- Concrete laying procedures

- Concrete pumping locations
- Finishing to the floor slab including the
  cutting of individual floor joints

## Risk assessments – site specific to concreting operations

**Risk Assessment Areas**

- Unloading and storage of materials

- Lifting and moving equipment manually

- Trowelling operations – mechanical operations

- Applying surface dust

- Working in poorly lit areas

- Moving and placing vehicles

- Placing concrete – concrete burns

- Manual handling

- Refuelling plant

- Working at heights

- Application of sealer sprays

- Waste management

- Securing small tools

- Concrete pumping

- Cutting mesh

- Use of electrical equipment

**Image of two storey steel frame to the first floor car park. Metal Deck formed of heavy duty HOLORIB metal deck supporting a 150 mm insitu concrete slab**

Metal decking being fixed in position to form six 18 m x 45 m floor bays to be concreted by mobile concrete pumps.

**General images of car park floor prior to commencing concreting**

## Concrete to large suspended floor

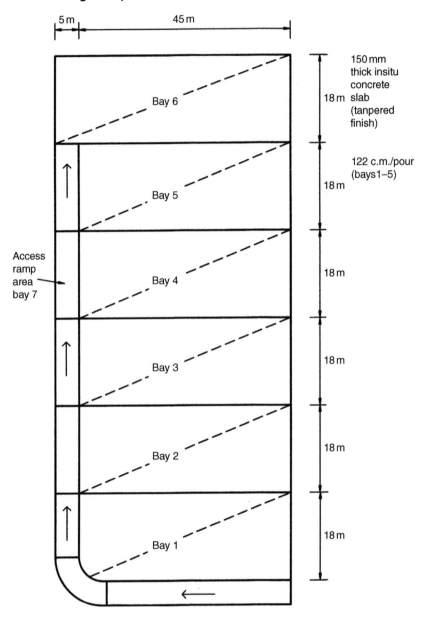

**Plan - 1st floor car park slab**

## Access for concrete pumping

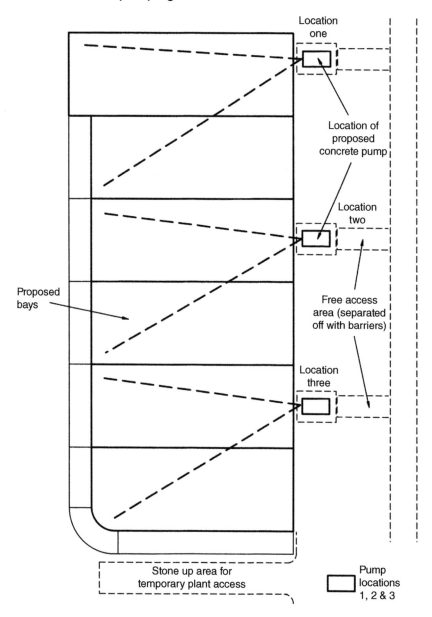

**Access for concrete pumping**

## Extract from contract method statement for concreting operations to suspended floor slab

Extracts from the subcontractors method statement covers the following areas:

### Concrete laying procedures to car park

The 150 mm concrete will be pumped in as indicated on the layout drawing. Levels to the car park to be given by the main contractor's engineer.

The concrete pump will be positioned at the pre-determined position and primed in readiness to accept the ready-mix concrete. The concrete will be laid utilising an aluminium straight edge based on a wet screed technique.

The ready-mix supplier has been briefed to ensure consistency, quality and reliability of the ready mixed concrete. The C28/35 concrete with added superplasticiser will be delivered to site at a rate of 36 m³ per hour.

### Concrete pumping considerations

Pump positions to be agreed prior to the works taking place, consideration should be given so that adequate space is provided to fully extend outriggers, unfold and fold boom, and not to be in the vicinity of any overhead power lines.

The main contractor to barrier off the pump position to prevent other site operatives entering the work area. The main contractor to ensure ground bearing pressure in pump position is suitable to support maximum rigger loads for a 24 m pump (127 kN per leg) in areas of bad ground. The main contractor may have to supply suitable spreader plates in order to achieve this.

Concrete must be suitably designed as a pumpable mix and quality must be monitored during the pumping operation. Delivery times must be managed to avoid concrete standing in the pipeline for excessive periods of time.

Safe manual handling techniques must be used when lifting pipelines that are full, it may be necessary for two operatives to handle pipelines, all concrete residue should be tipped out before final cleaning by the pump operator. A wash out area must be made available by the main contractor for the final clean down of the mobile pump.

### Where concrete is delivered via steel delivery pipes laid along the deck

- Where concrete is delivered via pipeline placed along the deck, ensure that all pipeline couplings are fitted with safety retaining clips.
- Ensure pipeline remains as straight as possible, where rubber pipes are used and ensure no kinks occur during the pumping operation.
- Only use a single ended coupler at the end of the delivery pipeline.
- Ensure operatives remain clear during initial priming of the pipeline.
- Pump driver is to act as a Banksman for reversing trucks onto the hopper. Readymix vehicles to have reversing horn fitted and operational during reversing procedures.

- During pumping operation the pump operator is to remain in a position where he can monitor the level of concrete in the hopper of the pump. If this is not possible, a nominated and competent person must take up the Banksman procedures and monitor concrete levels in the hopper and advise the pump operator as necessary.
- In the event of a blockage occurring, all operatives must retreat to a safe distance while the pump operator attempts to clear the line. If this is unsuccessful the pipe-line may have to be broken down and the blockage cleared manually, but must only be carried out on instruction from the pump operator.

## Finishing to car park

The finishing operations utilise both pedestrian powerfloats and will be undertaken by a dedicated two man team of experienced operators who will commence applying the brush finish operations approximately 4-6 hours after the initial placement and proceed down the length of the bay as the setting of the concrete takes place.

It is essential that the building is fully weathertight and that any openings are adequately covered to prevent the ingress of rain and drying winds.

On completion of the brushing operations the surface will be treated with a one-coat cure and sealing agent (Mastercure 181). Sawn induced joints will be cut within 48 hours of the concrete being cast (this will be dependent on the ambient temperature).

On all projects undertaken by the subcontractor, a working foreman is present at all times and will be made known to the site management. Each foreman is issued with a copy of the Method Statement and is responsible for ensuring its contents are communicated to the rest of his gang. The Contracts Manager will maintain communication with the site manager during his site visits.

All plant brought to site will be in good condition. Any faults, which arise, will be reported via the foreman to the office for immediate attention. All operatives will be trained in the correct use of all items of plant.

Each gang will have access to first aid facilities at all times. Each gang has access to a mobile phone, which can be used in the event of an accident occurring on-site when it is unattended (i.e. at night when the finishing of the concrete is still in progress).

All site operatives are issued with a full set of P.P.E., which must be used when conditions dictate. All welfare facilities are to be provided by the Principal Contractor.

## 12.9   Risk assessment – concrete operation

| HAZARD | PERSONNEL AT RISK | RISK CONTROL MEASURES | RISK ASSESSMENT | | |
|--------|-------------------|-----------------------|------|--------|-----|
| | | | HIGH | MEDIUM | LOW |
| Unloading and storage of materials | Site operatives other contractors | Only trained operatives to use fork lift truck<br>Use banksman where necessary<br>Store materials in allocated areas<br>Do not stack materials too high | | X | X<br>X<br>X |
| Trowelling operations | Site operatives | Operatives to use PPE equipment (boots, ear defenders)<br>Ensure guard rails in place | | X | X |
| Placing concrete Concrete burns | Site operatives | All operatives to wear suitable clothing - full PPE (must include gloves and eye protection)<br>eye wash to be available | | | X |
| Working at height | Site operatives | Work not to be commenced until suitable scaffold, ladders and access has been set up by main contractor | | X | |
| Application of sealer spray | Site operatives | Operatives to wear correct masks<br>masks and filters available to all operatives | | | X |

**Risk control measures**

## 12.10   Concrete placing to car park slab

**Ready mixed concrete supply vehicles feeding 120 cm floor pour**

**Mobile concrete pump located adjacent to bay pour. Pour commenced 8am – completed in 5/6 hour period**

**Setting up steel delivery pipes across deck to pouring location**

**Flexible delivery hose at discharge end of pipe line.**
**10 labourers engaged on placing reinforcement and laying concrete slab**

Screeding surface of slab with metal screed board.
Tolerances +/- or – 10 mm.
Tarmac surface to be laid over finished concrete

## Refer to subcontractors method statement and risk assessments

**Vast area of car park deck**

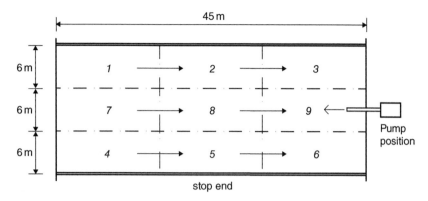

**Pouring sequence for a single bay pour. Pour areas 1 to 3, then 4 to 6, the final fill in 7 to 9**

## Completed two storey car park building – Opened July 2012

Completed car park building

**Roof to car park building**

## Completed supermarket building

**General view of laminated timber roof supported on 250 x 250 mm timber columns**

# Chapter 13

# University Refurbishment Project

## Contents

| | | |
|---|---|---|
| 13.1 | Contents of the case study | 252 |
| 13.2 | Project overview | 252 |
| 13.3 | Project information | 254 |
| 13.4 | Client/contractor relationships | 255 |
| 13.5 | Responsibilities of contractor's project team | 256 |
| 13.6 | Site management structure | 257 |
| 13.7 | Site layout plan and materials management | 258 |
| 13.8 | Sequence study – replacement of floor construction | 265 |
| 13.9 | Curtain wall panels to building elevations | 269 |

*Management of Construction Projects*, First Edition. Brian Cooke.
© 2015 John Wiley & Sons, Ltd. Published 2015 by John Wiley & Sons, Ltd.

## 13.1 Contents of the case study

**Architects impression of the completed Student Learning Centre**

This complex project involving both the removal and replacement of the insitu floors requires extensive co-ordination of work package contractors.

The case study encompasses the following aspects of both the management and construction methods encountered:

- An overview of the project information.
- The client/contractor relationships.
- The responsibilities of the contractors project team at senior management level.
- An overview of the contractor's site management.
- The contractor's approach to the management of materials during the contract.
- The integration of the frame refurbishment with the new insitu floor construction.
- The contractors approach to the external cladding of the building.

## 13.2 Project overview

The project involves the complete refurbishment of an existing three storey university building originally built in the 1960's period.

Carbonisation of the existing frame and floors requires the frame to be stripped back to its original column and beam structure. The insitu floors are to be removed and replaced with an insitu post tensioned troughed floor.

**Stripped back frame with new floor formwork in position**

The project is located on a major road in the centre of the University campus with difficult access problems. A new three storey insitu concrete frame is to be linked to the existing building at the rear of the complex.

**Building refurbishment and extension**

**Images of the "striped back" reinforced concrete frame**

Extensive siteworks involve a new landscaped garden area at the rear of the main building.

## 13.3  Project information

Client:                *Manchester University*
Contractor :           *Wates Construction*
Contract value:        *£12 million*
Contract period:       *13 months*
Form of Contract:      *Design and Build*
                       *(novated design team and single stage tender*

Architect:                          *Shepherd Robinson*
Project Managers:                   *A. A. Projects*
Structural Engineers:               *Gifford and Partners*
Landscape Architects:               *Gillespies*
Consultant Surveyors:               *Jacobs*
Building Services Consultant:  *RPS*

Work package contractors (up to 15):

- Demolitions
- Floor fabrications
- Concreting operations
- Brickwork/blockwork
- Steel work
- Internal lining
- Electrical services
- Heating and ventilation
- Curtain walling
- Roof decking
- External works
- Landscaping

## 13.4   Client/contractor relationships

**Design and Build Contract arrangement (with novation)**

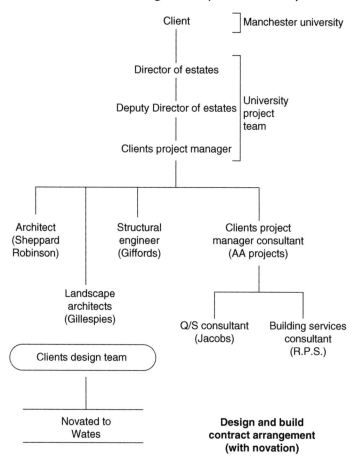

## 13.5 Responsibilities of contractor's project team

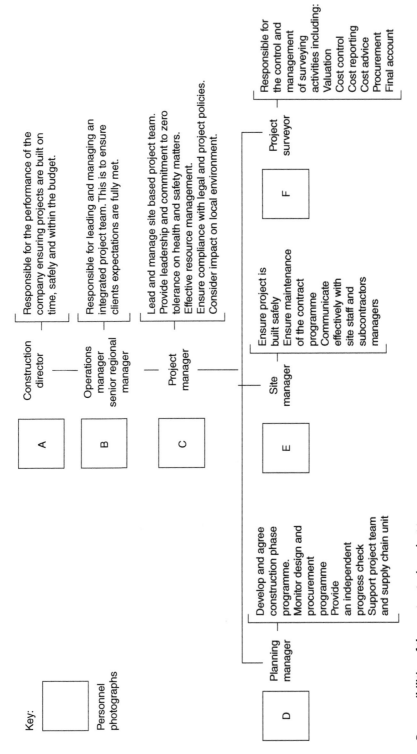

Key:

Personnel photographs

**A** Construction director — Responsible for the performance of the company ensuring projects are built on time, safely and within the budget.

**B** Operations manager senior regional manager — Responsible for leading and managing an integrated project team. This is to ensure clients expectations are fully met.

**C** Project manager — Lead and manage site based project team. Provide leadership and commitment to zero tolerance on health and safety matters. Effective resource management. Ensure compliance with legal and project policies. Consider impact on local environment.

**D** Planning manager — Develop and agree construction phase programme. Monitor design and procurement programme. Provide an independent progress check. Support project team and supply chain unit

**E** Site manager — Ensure project is built safely. Ensure maintenance of the contract programme. Communicate effectively with site staff and subcontractors managers

**F** Project surveyor — Responsible for the control and management of surveying activities including: Valuation. Cost control. Cost reporting. Cost advice. Procurement. Final account

**Responsibilities of the contractor's project team**

## 13.6   Site management structure

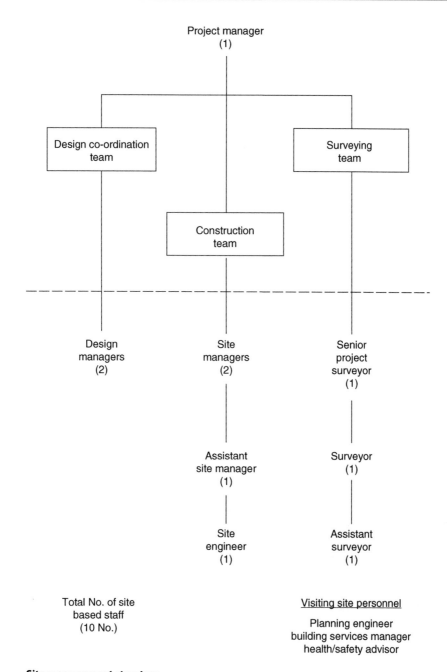

**Site management structure**

## 13.7   Site layout plan and materials management

### General site location

The site is located in the centre of a main university campus. Approximately £3 million is being expended on external works improvements outside the area of the site perimeter hoarding. This involves storing expensive paved materials (granite paving and polished sets etc.) within the main compound area.

The site layout plan indicates an extensive external works area to the south of the site accommodation which requires to be completed easily within the contract period.

### Site layout plan

The main contractor's compound area contains:

- A – Outdoor covered smoking area.
- B – Samples and mock panel area.
- C – Toilet block on ground floor and clients representative 1st floor.
- D – Main contractors accommodation on upper floor.
     Mess, canteen, drying room, induction room and reception on ground floor.

Adjacent area on site contains:

- E – Subcontractors offices.
- F – Paving and stone storage areas.
- G – Waste skip areas.
- H – General storage and work area in front of new frame extension.
- J – Fire points.

The site layout plan indicates pedestrian access along the north and east sides of the building. Access for crainage and hydraulic scissor lift platforms is required along all the elevations to facilitate the fixing of curtain wall support and glazed panels.

The main storage area in front of the new frame extension is to be used firstly, for the storage of the paving material and later for the glazed curtain wall panels. The site is located on one of the busiest main roads in Manchester. The site is surrounded by other university buildings currently occupied by students.

A dedicated site manager is responsible for the materials management. A name and shame notice board is used to notify subcontracts who do not comply with the site materials practices laid down by the main contractors.

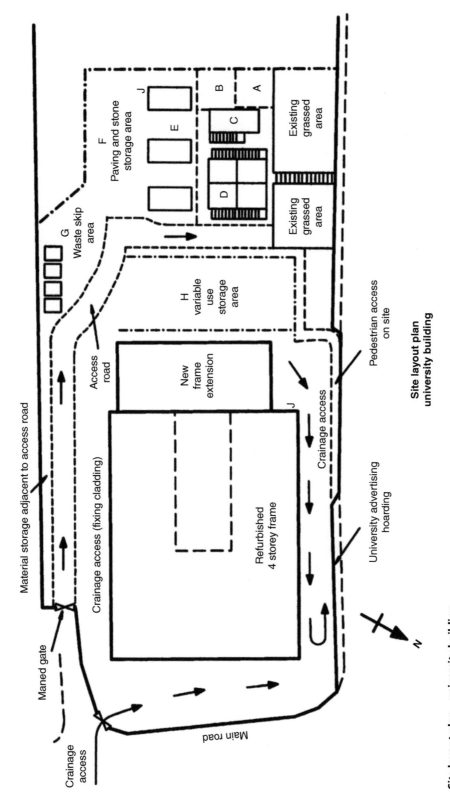

**Site layout plan – university building**

## Site layout plan and materials management

**Site office compound area – initial storage area for paving products in front of rear building extension**

**Storage adjacent to site access road**

Disbanded formwork support being banded and placed on storage racks prior to removal from site

Soffit timber being de-nailed and re-banded for transfer to another site. Loose timber skips for waste material

High security is necessary on the project due to the extensive circulation required for public and student access to the university buildings.

**Gated entry from site accommodation to site**

All deliveries to site are pre-scheduled due to the restrictive access for large vehicles from the road fronting the building.

**Steelwork storage area in front of building Storage adjacent to access road**

The site is fully protected by a high hording which clearly displays the purpose and function of the completed building.

Hazard board located adjacent to site entry and exit barrier.

**Site directional signs**

## Site management of materials

**Storage racks for vertical storage of curtain wall storage panels.**

**Cleared areas at each floor level for storage of internal wall lining materials**

Site Waste Management Plans incorporate the provision of dedicated material waste skips. Skips are marked for:

Mixed Waste:

- Packaging.
- Timber.
- Hardcore.

Wood Waste:

- Pallets.
- Offcuts.
- Plywood.
- Structural timber.

Metals Skip:

* Wire off cuts.
* Sealant and oil bin containers.

Oil storage platform for fuel drums.

**Skip storage area**

## 13.8   Sequence study – replacement of floor construction

**Description of work: method considerations in utilisation of the formwork**

**Formwork support**

- On completion of the removal of the insitu concrete floors by the demolition contractor, work can commence on constructing the new troughed floors between the existing floor beams.
- A decision has been made to cast each of the three floors and roof slab in two 100 cm pours per floor.
- This decision involves tying up expensive formwork support and soffits at any one time. The image above shows the formwork support in position for two floors of the building during the replacement of the floors.
- The ground floor slab formwork will be moved to the second floor position and on pouring the second floor and the first floor, formwork will be repositioned to the fourth floor/ roof level.

**The utilisation of the formwork**

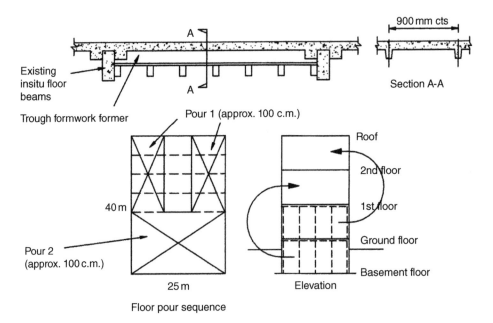

Existing insitu floor beams

Trough formwork former

Section A-A

900 mm cts

Pour 1 (approx. 100 c.m.)

40 m

Pour 2 (approx. 100 c.m.)

25 m

Roof

2nd floor

1st floor

Ground floor

Basement floor

Elevation

Floor pour sequence

**Method considerations in the utilisation of formwork**

## Formwork system

The images below show the troughed floor profiles being formed on the prepared soffit. The form of trough formed in the various floor areas was dependant upon the floor span. The longer spans incorporate post tensioning on completion of the placing of the concrete.

**Troughed floor profiles**

**Post - tensioning of troughs**

## Post tensioned floor – technology note

Post tensioning is a method of reinforcing concrete with high strength steel bars. In building, construction post tensioning allows longer clear spans and thinner floor slabs and allows a significant reduction in the weight of the floor beams.

The tendons are kept in a permanently stressed state which creates a compressive force to act on the concrete. This considerably increases the load carrying capacity of the concrete.

On completion of applying the load to the tendons the duct is filled with a cementitious grout that provides a corrosive protection to the tendons.

**Post tensioning tendons shown in position at each troughed beam**

**Floor slab poured in two 100 cm pours**

**Anchorages points on surface slab. Anchorage consists of iron casting and wedge system which grips the wire strand**

**View on underside of floor slab showing prestressed troughed beams**

## 13.9 Curtain wall panels to building elevations

**Rear elevation of building with aluminium curtain walling support in position to receive glazed panels**

**Aluminium framing bracketed to existing concrete frame at ground floor levels, floor edge beams and soffit to roof beams**

## Curtain wall panels – handling and fixing

**Handling and fixing of support frame and glazed wall panels from hydraulic scissor lift**

Components are handled onto the lift platform at ground floor level and raised to the fixing level. Clear glazing panels are fixed to complete the rear elevation of the building. Access is provided around the perimeter for plant and location of glass stillage platforms.

**Stillage platforms**

The side elevation of the building incorporates silver grey faced under cill panels to cover the existing wall up stand and floor edge.

**Elevation of rear external glazed wall**

Stillage platforms are used for storage of curtain wall panels and glazed units. Storage platforms are located around the building exterior.

**Image of side elevation of building**

# Chapter 14

# Managing a Small Business

## Contents

| | | |
|---|---|---|
| 14.1 | Contents of the case study | 273 |
| 14.2 | Managing a small building enterprise | 274 |
| 14.3 | Business operations | 274 |
| 14.4 | Current and completed developments | 277 |
| 14.5 | Project case study | 278 |
| 14.6 | Around the site | 281 |
| 14.7 | Foundations and ground floor slab to timber framed house units | 283 |
| 14.8 | Site fabrication process | 285 |
| 14.9 | External elevations | 288 |
| 14.10 | Erection sequence for ground floor panels | 289 |
| 14.11 | Comments on the organisation and management of business | 292 |

*Management of Construction Projects*, First Edition. Brian Cooke.
© 2015 John Wiley & Sons, Ltd. Published 2015 by John Wiley & Sons, Ltd.

## 14.1   Contents of the case study

**Evidence of a well planned project**

The establishment of a small building firm is outlined in this chapter. The company described, undertakes construction work for two property development clients. The managing director's background experience is outlined and his approach to managing a small housing contract clearly shown.

The following aspects are covered:

- The responsibilities of the principal director and his quantity surveyor/ office administrator are outlined.
- Images of the office facilities for a small contractor are shown.
- The approach to establishing the site facilities for a development of nine housing units is illustrated. A site layout plan is shown for the project.
- The site strategy is outlined for the project in relation to the planned building sequence and site layout plan.
- The sequence of construction for a semi-detached timber framed unit is illustrated. The images are shown of the foundations and ground floor slab construction. The

various stages of the erection of the timber frame panels are shown including a detailed sequence of erecting the ground floor panels.

## 14.2 Managing a small building enterprise

Contact was made with the managing director/ owner of a small building firm. The company, G. Construction are specialists in building small speculative housing projects for two developer clients. Projects completed to date vary from two house units to a maximum of ten houses on a single project.

### Company background

The business was registered as a limited liability company in 2009. The managing director has ten years previous experience as a Contracts Manager with a north west developer. The previous company went into receivership in 2008.

The principal of G. Construction has to date developed a good working relationship with two developer clients. The developers are willing to fund new housing projects and allow G. Construction to build them.

G. Construction are to be responsible for organising the build work using subcontract trades that they have used extensively in the previous company. An excellent relationship based on trust and fairness has been established with them and they are considered an essential part of the contractor's team.

The developer clients are responsible for all planning decisions, design, site services fees and building regulations approvals.

The developer and contractor are responsible for agreeing an operational budget for each stage of the subcontract work and establishing an overall cost forecast. It has been agreed that any saving on the operational budget is to be shared equally between the developer and the contractor.

The contract between the developer and the contractor (G. Construction) is based on the principles of the JCT cost plus contract arrangement. The contractors monthly expenditure claim (the cost) is based on invoiced costs submitted to the developer plus a 10% addition.

The turnover of G. Construction has increased from £100,000 to £900,000 over the past five years and the company profit margin is in the range of 8% – 12%.

## 14.3 Business operations

The company operate from rented office accommodation within the north west region of England. Images of the office environment created by the contractor are shown.

## Contractor's office

**Well organised office showing good use of wall boards and filing systems**

**Computers and office photo-copying facilities. All office systems and routines are computerised**

**Directors office space showing use of wall boards to highlight key daily and weekly tasks**

The organisation structure for the company is illustrated below, indicating the responsibilities of each member of the team. These individuals make up the total staff.

| **Principal Director** | *Company policy* <br> *Contact with clients* <br> *Direct site management of subcontractors* <br> *Co-ordination of site operations* <br> *Verifying accounts to clients* <br> *Negotiations with clients* |
| --- | --- |
| **Quantity Surveyor/ Office administrator** | *Office administration* <br> *Scheduling quantities and preparing material schedules* <br> *Payments to subcontractors* <br> *Materials scheduling* <br> *Monthly valuations where appropriate* <br> *Dealing with subcontract and tender package enquiries* |
| **Secretary (part time)** | *Filing – letters and accounts* <br> *Payment records* <br> *Correspondence* <br> *Petty cash accounts* |

Over the past five years, five sites have been developed. These have varied from a single dwelling site up to a development of eight to twelve houses. A site layout plan for a nine house development is shown together with a view of a completed development.

## 14.4 Current and completed developments

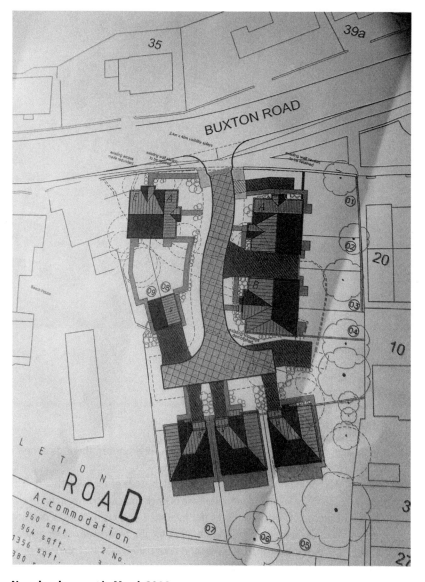

**New development in March 2014**

**Completed development in 2013**

## 14.5   Project case study

### Managing the Buxton Road project

The development includes the construction of nine houses, three detached units and six semi-detached houses.

The project commenced in September 2013 at the start of a so called housing boom. The housing units were intended to be made of brick and block construction, however, due to a shortage of concrete blocks (for internal wall construction), a decision has been taken to use timber frame construction for the inner walls, floor and roof. It has been decided to construct the wall and floor panels on site. Engineers drawings have been prepared to aid this task (refer to the fabrication area on the site layout plan).

A plan of the scheme is shown together with proposals for the site layout planning.

**Layout plan – Buxton Road**

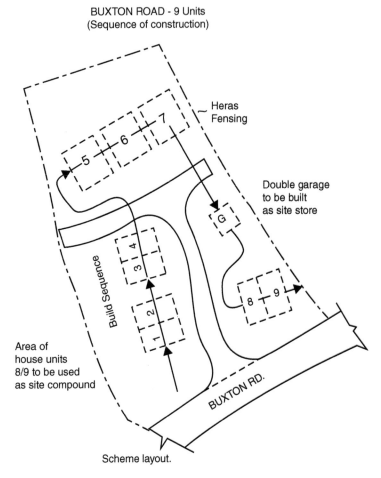

BUXTON ROAD - 9 Units
(Sequence of construction)

Heras
Fensing

Double garage
to be built
as site store

Build Sequence

Area of
house units
8/9 to be used
as site compound

BUXTON RD.

Scheme layout.

**Buxton Road – 9 units (sequence of construction)**

## Site strategy

- The build sequence is shown for houses 1 to 9.
- The access road is to be constructed to base course level prior to commencing the foundations work.
- A double garage (marked block G) to be constructed in order to be used as a secure lock up store for the project.
- Ground floor slabs for units 5, 6 and 7 to be completed. Block 7 is to be used as a fabrication area for the site manufacture of the timber frame wall and floor panels. Completed panels to be stored on block 5 and 6 prior to erection by safe handler.
- Temporary site office and plant and materials store (marked 1 and 2 on the layout plan) to be provided together with attached toilet facilities.
- Completed timber framed panels to be stored on the ground floor slab areas of blocks 5 and 6 prior to erection.

- Area at side of garage G to be used for a general material storage area.
- Site manufactured timber frame panels to be stacked on floor slab of each block prior to erection by telehandler/ lifter, assisted by manual labour.

A quotation for the supply and erection of the completed timber frames to block 1 and 2 was obtained to the sum of approximately £26,000. The contractor considered that the cost of site manufacture plus erection would be in the order of some £4,000 – £5,000 or less. Overall this would result in a saving on the project of some £25,000 – £27,000.

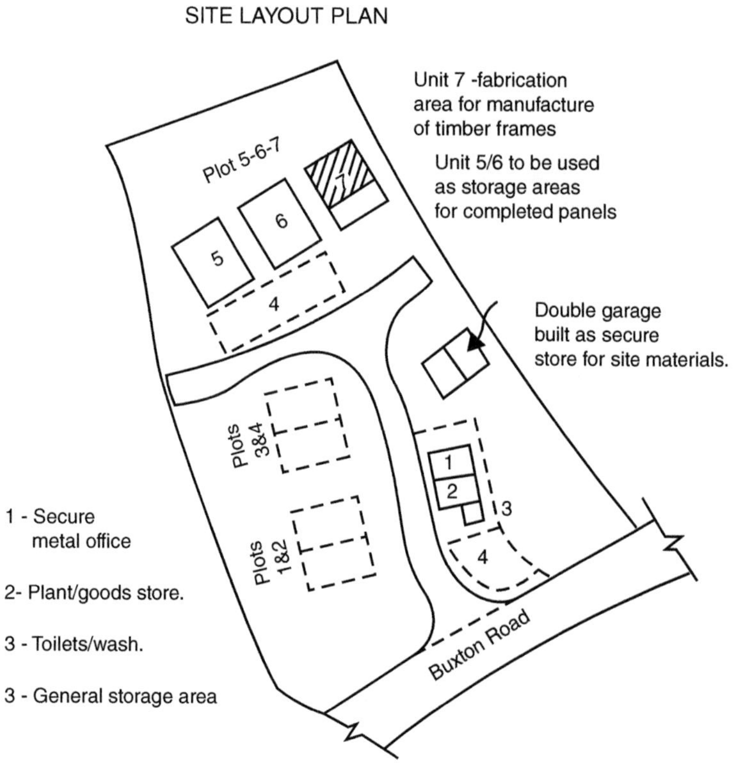

SITE LAYOUT PLAN

Site layout plan

# 14.6 Around the site

## Offices and storage facilities

The compound and area around the offices to be stored up and secure metal cabins provided for the office and plant store.

**View of site office facility**

The single unit office is to accommodate the site manager, provide office facilities and house the water and power supply.

**Microwave facility for staff and operatives**

A double garage to be built in plot 8 and 9, to accommodate plastic manhole courses, ironmongery items and internal doors etc. Components being stored will vary during the contract.

**View of double garage facility**

## 14.7 Foundations and ground floor slab to timber framed house units

**Foundations and ground floor slab**

One metre deep strip foundations constructed of heavy trench blocks (300 mm wide x 200 mm deep) and ten courses of engineering brick up to the top of the ground floor slab level.

The ground floor to be constructed of precast beam and pot floor construction, incorporating methane gas membrane.

**Beam and block ground floor slab**

**Sole plates for wall panels 150 mm x 40 mm**

## 14.8 Site fabrication process

Timber sole plates to all internal and external wall panels. The gas membrane is incorporated into the floor construction. Timber wall panels 150 mm wide and shot fired fixing of panels to timber sole plates.

**Site fabricated 150 mm wide internal and party wall panels**

The panels are manufactured in the site fabrication area. The panels are handled with a hydraulic lifter onto the floor slab area.

**Erection of ground floor panels to plot 1**

The first floor panels are put into position, incorporating IJ floor beams (see internal images).

The wall panels are erected by two joiners and one labourer within a one day period.

## Internal frame and floor panels

**On site panel fabrication area- scaffold framed temporary building with open front**

**General internal view of ground floor area**

**Internal timber frame and floors**

**Underside of floor panels to first floor – IJ beams at centres incorporated in floor construction**

## 14.9 External elevations

**Front and side elevation panels to upper floor in position. Upper floor panels to house units 1 and 2 erected within a one day period (two joiners and one labourer)**

**Wall panels to upper floor of semi-detached frame**

## 14.10 Erection sequence for ground floor panels

**Commencement of erection of panels at ground floor level**

Site fabricated timber panels positioned on ground floor slab by an hydraulic lifter. Wall panels handled and fixed in position within a four hour period by two joiners, one labourer and the site manager.

Eighteen wall panels make up the complete ground floor plan of a semi-detached house unit.

## 14.11 Comments on the organisation and management of business

- The principal/ director of this housebuilding company has – "GOT IT RIGHT!"
- The business has reliable clients, who pay promptly each month and a good working relationship exists between them.
- A high degree of trust exists between the client and the contractor and future development projects are being jointly worked on. This should ensure continuity of work for the business.
- At the current time, the private/ speculative housing market is buoyant and encouraged by the government's – "help to buy scheme" and the banks releasing more mortgages.
- The principal/ director is directly involved in the site management of his projects. A stable subcontractor labour force is employed which will need to be expanded as additional projects commence.
- As further developments come on stream, the contractor will have to consider recruiting site management personnel (i.e. additional site managers).
- Turnover currently is in the order of £100,000 per month, over expansion of the company will have to be carefully monitored.

# Index

Asta developments plc, 35
  refer to Power Project examples,
    36–8

client/contractor relationships
  Chinley School project, 197
  co-operative building, 168
  Merlin project, 146
  retail unit and car park, 224
  university refurbishment, 255
contract JCT family
  construction management contract
    2011, 53
  contractor led/client led
    relationships diagrams, 48–9
  design and build contract 2011,
    49–56
  intermediate building contract
    2011, 47
  major projects construction contract
    2011, 47
  management building contract
    2011, 53
  standard building contract 2011, 46
crane selection factors, 74
crane types, 75

defects
  managing, 95
  recording, 96
  stages, 103–4
display boards
  daily hazard board, 13
  daily task board, 13
  environmental display board, 11
  name and shame board, 14
  site layout plan, 11
  site progress photographs, 12

functional relationships and line
    management, 5

linked bar chart programmes
  preparing a linked bar chart, 36–8
  principles and relationships, 39
  procurement programme
    extract, 43
  procurement programme principles,
    40–44
  procurement symbols, 42
  programme extracts, 218–20
logistic examples
  concrete pump locations, 182
  fire and traffic plans, 181
  site layout proposals, 228
  site tower crane locations, 179
loss and waste, site planning to
    reduce, 56

materials management
  bad practice observations, 56
  Chinley School project, 209
  good practice observations, 65–9
  Merlin project, 150
  retail unit and car park
    project, 229
  site planning to reduce loss and
    waste, 56
  university refurbishment project,
    258–64

*Management of Construction Projects*, First Edition. Brian Cooke.
© 2015 John Wiley & Sons, Ltd. Published 2015 by John Wiley & Sons, Ltd.

organisation structures
organisation principles, 4
regional organisation, 7

plant types
aerial work platform, 81
crane types, 75
hydraulic lifter, 84
telescopic handler, 79
programme techniques, overview, 35

qualifications for site management, 8

records maintained on the site
site instructions/variations, 204
site visitors, 203
site waste management plan, 209–13
weather, 203
weekly site report, 204–5
monthly progress report, 206
subcontractor's meetings, 207–8
regional site organisation, 7
risk assessment areas, 87
risk assessment examples
access and egress, 88
cherry picker, 86
concreting operations, 246
delivery of materials, 89
heavy plant and machinery, 78
mobile elevated platform, 83
personal equipment (PPE), 90
schedule of risk assessments, 87
roles and responsibilities of
the clerk of works, 21
the planning engineer, 15
the procurement manager, 16
the project manager, 9
the project surveyor, 16
the site engineer, 18
site management personnel, 6–7
the site manager, 14

sequence of erection studies
curtain wall panels, 269–71
external storey height cladding
panels, 189–94
main roof construction,
213–17
multi-storey steel frame,
183–9
replacement of concrete floors,
265–8
timber glulam frame erection
sequence, 232–7
site layout plans
Buxton Road project, 281
Chinley School project,
199–200
co-op building logistics, 178
hotel and office project, 118
retail unit and car park, 228
university refurbishment
project, 258
site organisation structure
Chinley School, 198
co-operative Project, 105–6
hotel and office project, 112
Merlin project, 148
retail unit and car park, 227
university refurbishment
project, 257

team building
developing teams, 24–9
procurement team, 26
site engineering team, 29
site management team, 25
surveying team (quantity
surveyors), 27
tender negotiations with the
preferred bidder, 112

value engineering proposals, 114

Printed and bound by CPI Group (UK) Ltd, Croydon, CR0 4YY

16/04/2025

14658834-0001